Synthesis Lectures on Digital Circuits & Systems

Series Editor

Mitchell A. Thornton, Southern Methodist University, Dallas, USA

This series includes titles of interest to students, professionals, and researchers in the area of design and analysis of digital circuits and systems. Each Lecture is self-contained and focuses on the background information required to understand the subject matter and practical case studies that illustrate applications. The format of a Lecture is structured such that each will be devoted to a specific topic in digital circuits and systems rather than a larger overview of several topics such as that found in a comprehensive handbook. The Lectures cover both well-established areas as well as newly developed or emerging material in digital circuits and systems design and analysis.

Preface

Texas Instruments is well known for its analog and digital devices, in particular, Digital Signal Processing (DSP) chips. Unknown to many, the company quietly developed its microcontroller division in the early 1990s and started producing a family of controllers aimed mainly for embedded meter applications for power companies, which require an extended operating time without recharging. It was not until the mid-2000s, the company began to put serious efforts to present the MSP430 microcontroller family, its flagship microcontrollers, to the academic community and future engineers. Their efforts have been attracting many educators and students due to the MSP430's cost and the suitability of the controller for capstone design projects requiring microcontrollers. In addition, Texas Instruments offers a large number of compatible analog and digital devices that can expand the range of the possible embedded applications of the microcontroller.

Background

This manual provides a thorough, practical laboratory introduction to the Texas Instruments MSP430^{TM} microcontroller. The MSP430 is a 16-bit reduced instruction set (RISC) processor that features ultra-low power consumption and integrated digital and analog hardware. Variants of the MSP430 microcontroller have been in production since 1993. This provides for a host of MSP430 products including evaluation boards, compilers, software examples, and documentation. All labs use the Texas Instruments MSP-EXP430G2ET LaunchPad as shown in Fig. 1 which uses the MSP430G2553 microprocessor. This LaunchPad was introduced in June 2018 and is the replacement for the MSP-EXP430G2 LaunchPad. The labs have been designed to introduce you to the basic functionality of the MSP430G2553 microprocessor. The manual features use of the Digilent Analog Discovery 2. However, other laboratory equipment may be used. The manual is intended for an upper level undergraduate course in microcontrollers or mechatronics but may also be used as a reference for capstone design projects. Also, practicing engineers already familiar with another microcontroller, who require a quick hands-on tutorial on the microcontroller, will find this book very useful.

Fig. 1 Texas Instruments MSP-EXP430G2ET LaunchPad

Approach and Organization

The labs are designed to introduce you to the basic workings of the MSP430G2553 micro-controller. You will be programming the microcontroller using code listings provided with the labs in both the C programming language and the MSP430 Assembly Language.

Many times not everything works as planned and troubleshooting is necessary to deter-mine where the problem may be. Things to consider may be (1) Was the code loaded properly? (2) Are there errors in the code? (3) Are the jumper wires in the right place? and

(4) are all ground connections in place? An important skill for an engineer is developing good troubleshooting ability.

The code contained in this manual was designed and intended as additional material to help introduce basic applications using the MSP430 microcontroller. Detailed explanations are provided with the actual code to help understand program processes. The code listings are not optimized for efficiency, or speed and were designed for learning purposes. The code can be modified and used for other applications. It is the responsibility of the end user to ensure particular applications have been thoroughly tested.

Synopsis of Labs

1. Introduction to the MSP-EXP430G2ET LaunchPad and ENERGIA. Energia is an open-source platform that is free and can be downloaded from energia.nu. When you first run Energia it will most likely require some setup to communicate with the MSP-EXP430G2ET LaunchPad.

2. Introduction to Texas Instruments Code Composer IDE. In this lab, you will receive your first exposure to programming the MSP430 Microprocessor (MSP430G2553) using Texas Instruments Code Composer IDE. It allows the user to begin with a "framework" program for C or Assembly code. As you get more familiar with using Code Composer you will find that it has features to examine what is stored in registers and addresses which makes finding problems in code easier.

3. Introduction to the General Purpose Digital I/O pins. In this lab, you will look in depth at the input/output capabilities of the MSP430G2553. You will learn how to declare an I/O pin as an input or output within Port 1 (8 pins available) and Port 2 (8 pins available).

4. Exploring the Basic Clock Module of the MSP430G2553. At the heart of the microcontroller CPU is the clock system. The MSP430G2553 allows for several different clock speeds and configurations. In later labs, you will see how the clock system is important in determining timing parameters.

5. Interfacing a 16x2 Liquid Crystal Display (LCD) with the MSP430G2553. Microcontrollers can do many things; however, it is nice to be able to visually see the output of a sensor or other device. In this lab, you will learn how to interface a common 16x2 LCD display, and write code to maneuver around the screen.

6. Introduction to the Analog-to-Digital Converter (ADC) Module on the MSP430G2553. In this lab, you will learn how to use the ADC module on the MSP430G2553. This will allow you to incorporate sensors and other devices into your projects. Using your skills learned in Lab 5, you will be able to see the output on an LCD Display.

7. Introduction to the Timer_A Module of the MSP430G2553 (Introduction to Interrupts). Timer modules are an important part of a microprocessor. In this lab, you will gain experience using the timer module and in the process also learn about interrupts.

8. Introduction to the Timer_A Module of the MSP430G2553 (Introduction to Pulse Width Modulation (PWM). This lab is a continuation of learning about the timer module. You will use the timer module to generate a PWM signal to control the brightness of a Light Emitting Diode (LED). Changing timing parameters you will visually see the different outcomes.

9. Introduction to the Universal Asynchronous Receiver/Transmitter Serial Communication Interface (UART Mode) Module of the MSP430G2553. Getting data into and out of the MSP430G2553 is done in many ways. Built into the microcontroller are modules that can help with data communication. One of these is the UART serial format module. In this lab, you will be using two microcontrollers, one that will transmit data, and one that will receive data and output it on a LCD Display.

Synopsis of Helpful Reference Documents

Here is a list of helpful reference documents. Many of these are available in the Appendices. Others are available on manufacturers' websites.

The following information is available in the Appendices:

- Introduction Digilent Analog Discovery 2. This appendix provides the basics on how to use the Digilent Analog Discovery 2 (DAD2) PC-based test instrument. The DAD2, along with a host personal computer (PC) or laptop, provides a full suite of portable test equipment.
- Header File for the MSP430G2553. You will find this document very useful because many registers and pins can be addressed by different names and numbers giving the same results. When confusion arises (and it will), look it up in the Header file.
- DEBUG in Code Composer (How to view what is in the registers). This document will guide you through how to look at what is happening in the registers of the MSP430G2553 when using Code Composer. It presents just the basics.
- Controlling a 16x2 LCD Display (Useful information for Labs 5,6,7,9). This document provides information on controlling a 16x2 LCD Display with details in an understandable format with pictures.
- Useful Miscellaneous Information (Tidbits of useful information). In the course of preparing the labs for this manual several bits and pieces of information were "discovered." Some of these glitches took several hours to discover and figure out. This document is a compilation of these tidbits of information that may save you from spending hours trying to figure out why something isn't working out as planned.
- Summary of C Operators. This document will be useful when programming in the C language and explains the syntax for different operations.
- A guideline to developing Unified Modeling Language (UML) activity diagrams (flowcharts).

Other information is available from manufacturers' websites:

- Texas Instruments Document SLAS735J (Datasheet for MSP430G2553). This is the datasheet for the MSP430G2553 microcontroller and will be very useful when working with the features that are unique to the microcontroller. It will be very useful when working with the secondary functions of microcontroller pins (www.TI.Com).
- Texas Instruments Document SLAU144K (MSP430 Users Guide). This document is the generic Users Guide for the MSP430 family of microcontrollers. It describes all of the modules in detail. Not all microcontrollers have all of the modules and capabilities that are presented. You will find this document very useful (www.TI.Com).
- Newhaven LCD Display Datasheet. The 16x2 LCD Display is very common, and they come in a wide variety of styles (letter colors, background colors, etc.). Common to most is the HD44780 driver (or newer ST7066U00) (newhavendisplay.com). This document is rather technical, see "Controlling a 16x2 LCD Display."

St. George, USA   James Kretzschmar
St. George, USA   Jeffrey Anderson
Laramie, USA   Steven F. Barrett

Acknowledgments

There have been many people involved in the conception and production of this book.

Debbie Kretzschmar wrote the C-Code program for driving the LCD display which appears in the Appendices. The detailed explanations of the functions provide a thorough understanding of the processes involved in making the LCD display operate for intended applications. Debbie has a B.S. degree in Computer Science from Oregon State University, and after a career in the U.S. Air Force has been involved with the Electrical Engineering department at the University of Wyoming as a Teaching Assistant, and with summer outreach programs for high school students. She is also an amateur radio operator (FCC license KF7CJF).

Dave Feldman provided extensive help with troubleshooting the UART_RECEIVER and the UART_TRANSMITTER communication Assembly programs. Dave has a M.S. degree in Electrical Engineering from the University of Colorado at Denver, and after a career in network technologies continues to pursue his interest in microcontroller real-world applications. Dave has been an amateur radio operator (FCC license WB0GAZ) for over 50 years and is currently developing a microcontroller-based UHF/microwave transceiver.

The appendix on how to interface a common 16x2 LCD Display to the Texas Instruments MSP430G2553 microcontroller originally appeared in "Controlling a 16x2 LCD with the Texas Instruments MSP430G2553 Microcontroller," QEX-A Forum for Communications Experimenters, No.199 Nov/Dec 2021. We gratefully acknowledge the American Radio Relay League's (ARRL), www.arrl.org, for permission to include it in this book.

We would also like to acknowledge Joel Claypool for his publishing expertise and support on a number of writing projects. His vision and expertise in the publishing world has made this book possible. Joel "retired" in September 2022 after 40 plus years of service to the U.S. Navy and the publishing world. On behalf of the multitude of writers you have provided a chance to become published authors, we thank you!

We would also like to thank Charles (Chuck) Glaser, Editorial Director at Springer Nature, for his encouragement and support on this project. If you have a good idea for a book, we highly recommend contacting Chuck. He will assist you in converting your idea into a finished, professional book product.

We would also like to thank Dharaneeswaran Sundaramurthy of Total Service Books Production for his expertise in converting the final draft into a finished product. You provide outstanding service.

Finally, we thank our families who have provided their ongoing support and understanding while we worked on the book.

St. George, UT, USA   James Kretzschmar
St. George, UT, USA   Jeffrey Anderson
Laramie, WY, USA   Steven F. Barrett

Contents

Lab 1: MSP–EXP430G2ET LaunchPad and Energia

1

1.1 Purpose

In this lab you receive your first exposure to programming a MSP430 microcontroller (the MSP430G2553 20–pin chip included with the MSP–EXP430G2ET LaunchPad). Energia is a free integrated development environment (IDE) that provides a C/C++ coding platform, and a user–friendly interface for uploading code into the microcontroller (MSP430G2553). As part of this lab an oscilloscope is used to observe the timing of events happening within the microcontroller as it executes the uploaded code.

1.2 Energia IDE

Energia is an open–source platform that is free and can be downloaded to your computer. Visit https://energia.nu to down load the latest version. When Energia is run for the first time it will most likely require some setup to communicate with the MSP–EXP430G2ET LaunchPad. The opening screen should look like Fig. 1.1.

The first order of business is to make sure that the Energia IDE can communicate with the MSP–EXP430G2ET LaunchPad. Do the following:

1. Plug in the LaunchPad into a USB port on the computer.
2. Select "Tools" and go to "Board" and it should read "MSP–EXP430G2ET w/ MSP430G2553." If it does not then it will be necessary to go through the process of installing the parts that allow communication from the IDE to the proper Texas Instruments development board (MSP–EXP430G2ET) and microcontroller (MSP430G2553). If this step is needed, select the "MSP–EXP430G2ET" board and follow the installation process.

© The Author(s), under exclusive license to Springer Nature Switzerland AG 2023
J. Kretzschmar et al., *MSP430 Microcontroller Lab Manual*, Synthesis Lectures on Digital
Circuits & Systems, https://doi.org/10.1007/978-3-031-26643-0_1

Fig. 1.1 Energia IDE

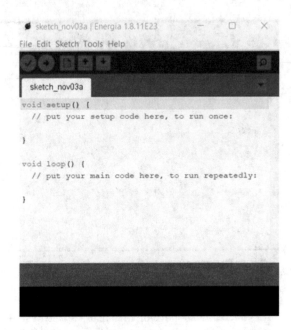

3. Select "Tools" and go to "Port" (a port such as COM3 should be listed). If there are multiple Ports listed it may be necessary to move the LaunchPad to another USB port. If there are no ports displayed, ensure that the LaunchPad is securely connected to the computer, and that the USB cable is firmly in place. If no connection, it may be necessary to look at the COM ports in the device manager screen of the computer and adjust on the computer.

1.3   Getting Started

There are several example programs (called Sketches) under the "File" menu button. Select "File," "Examples," "01.Basics," and "Blink." A new screen appears with the "Blink" program (Sketch) in place (see Fig. 1.2). The section labeled "void setup" sets up parameters such as defining pins on the microcontroller as input pins, or output pins. Setting up these parameters only needs to be accomplished one time. All of these statements are contained within the two brackets. The next section which is labeled "void loop" contains statements instructing the microcontroller to accomplish a task such as blink a LED, or input data from a sensor. All of these statements are also contained within the two brackets.

The double hash marks "//" mean that what follows is not executed. These hash marks are used to make a comment (describe what is happening in the program), or temporarily keep a statement from being executed. The statement "# define LED GREEN_LED" declares that pin P1.0 (pin #2 on the microcontroller) has a green LED attached to it. Similarly, the "#

File Edit Sketch Tools Help

Blink

```
Blink
The basic Energia example.
Turns on an LED on for one second, then off for one second, repeatedly.
Change the LED define to blink other LEDs.

Hardware Required:
* LaunchPad with an LED

This example code is in the public domain.
*/

// most launchpads have a red LED
#define LED RED_LED

//see pins_energia.h for more LED definitions
//#define LED GREEN_LED

// the setup routine runs once when you press reset:
void setup() {
  // initialize the digital pin as an output.
  pinMode(LED, OUTPUT);
}

// the loop routine runs over and over again forever:
void loop() {
  digitalWrite(LED, HIGH);    // turn the LED on (HIGH is the voltage level)
  delay(1000);                // wait for a second
  digitalWrite(LED, LOW);     // turn the LED off by making the voltage LOW
  delay(1000);                // wait for a second
}
```

Fig. 1.2 Energia IDE blink sketch

define LED RED_LED" statement declares that pin P1.6 (pin #14) has a red LED attached to it. Note the "//" in front of this statement (this statement is not executed). These two define statements are specific for the MSP–EXP430G2ET LaunchPad because there are red and green LEDs that are wired to pin P1.0 and pin P1.6.

The statement "digitalWrite (LED, HIGH)" tells the microcontroller to turn on the green LED (+3.3 V is placed on pin P1.0). The statement "delay(1000)" tells the microcontroller to wait for 1,000 ms before executing the next line of code. The statement "digitalWrite (LED, LOW)" now tells the microcontroller to turn off the green LED (0 V placed on pin P1.0). The next statement is "delay(1000)" again. The program now goes back to the beginning of "void loop" repeat the process over and over.

Selecting the right–pointing arrow in the circle at the top of screen uploads the program into the microcontroller and begins execution of the program. Immediately the green LED should begin to blink. Selecting this button sets in motion the compiling, assembling, linking, and uploading of the program into the microcontroller. All of these processes are taking place in the background within the IDE.

1.4   Further Exploration

Once the "Blink" program is running make the following adjustments in the program and observe the results.

1. Turn on the green LED for 0.2 s; change to delay (200).
2. Turn off the green for 0.5 s; change to delay (500).

   Connect an oscilloscope to pin P1.0 (pin #2 of the microcontroller) and look at the timing pattern. Does the timing pattern correlate with your program code? To save your program go to "File," "Save As," and give the program a name. The example code that you started with is a read only file. The oscilloscope signal trace should look similar to Fig. 1.3.

   Adjust the timing of the on/off pattern of the green LED to create a message in Morse Code using the following code listing. Use the "Blink" example as a guide to complete the program. Morse Code was developed by Samuel Morse in the mid–1800s and although not used much today the international signal for distress "SOS" is still recognized throughout the world. The letter "S" is three short "ON" pulses, and the letter "O" is three long "ON" pulses as shown in Fig. 1.4.

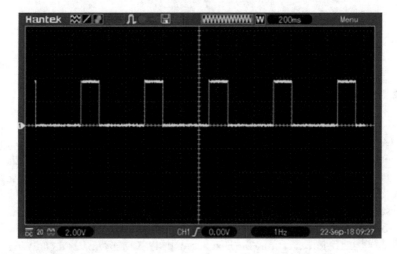

Fig. 1.3 Timing pattern

```
//******************************************************************
// Morse Code ''SOS'' Program:
//******************************************************************

digitalWrite(LED, HIGH); // First pulse of the letter 'S'
delay(150);
digitalWrite(LED, LOW);
delay(400);
digitalWrite(LED, HIGH); // Second pulse of the letter 'S'
delay(150);
digitalWrite(LED, LOW);
delay(400);
digitalWrite(LED, HIGH); // Third pulse of the letter 'S'
delay(150);
digitalWrite(LED, LOW);

delay(1000); // Space between letters

digitalWrite(LED, HIGH); // First pulse of the letter 'O'
delay(750);
digitalWrite(LED, LOW);
delay(400);
digitalWrite(LED, HIGH); // Second pulse of the letter 'O'
delay(750);
digitalWrite(LED, LOW);
delay(400);
digitalWrite(LED, HIGH); // Third pulse of the letter 'O'
delay(750);
digitalWrite(LED, LOW);

delay(1000); // Space between letters

digitalWrite(LED, HIGH); // First pulse of the letter 'S'
delay(150);
digitalWrite(LED, LOW);
delay(400);
digitalWrite(LED, HIGH); // Second pulse of the letter 'S'
delay(150);
digitalWrite(LED, LOW);
delay(400);
digitalWrite(LED, HIGH); // Third pulse of the letter 'S'
delay(150);
digitalWrite(LED, LOW);

delay(3000); // Space between multiple 'SOS' signals
//******************************************************************
```

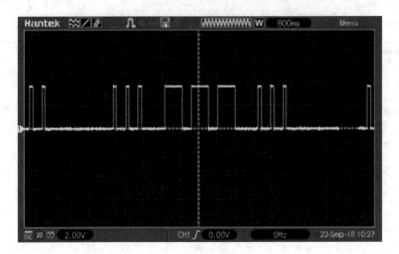

Fig. 1.4 SOS timing pattern

The Morse Code "SOS," program, or any Morse Code program can easily be adapted to control the on/off of a radio transmitter. Some licensed amateur radio operators use micro-controllers such as the MSP430G2553 to automatically send signals on radio frequencies where it is permitted to have an unattended radio transmitter.

1.5   Post Lab Activities

1. In the Appendices we provide a brief introduction to Unified Modeling Language (UML) activity diagrams. These are UML compliant flow charts. Provide UML activity diagrams for programs provided in this laboratory exercise.
2. For assembly language programs provided; rewrite in C, compile, execute, and test for proper operation.

Lab 2: Introduction to TI Code Composer IDE

2

2.1 Purpose

In this lab you receive your first exposure to programming a MSP430 microcontroller (the MSP430G2553 20–pin chip included with the LaunchPad) by using the TI Code Composer IDE. For purposes of getting familiar with Code Composer we write two very simple programs, one in assembly code, and the other in C code and running them on the MSP430G2553 microcontroller. Then in Task C you write an assembly program that sequences three LEDs in an on/off pattern. The main purpose of this lab is to introduce you to Code Composer and how to maneuver around the IDE to program the microcontroller.

2.2 Getting Started Guide

The latest version of Code Composer is available for free from the Texas Instruments website (www.TI.com) to load on your personal computer. Once you have registered with TI you are able to complete the download. Because Code Composer is a large program it takes a while to download, and you may need to temporarily turn off some of your security protection as you download the program so it does not hit a snag.

2.3 Prelab

2.3.1 Task A

1. Read through the documentation for this lab so you are well prepared for the other Tasks.
2. At the top of programs you see "msp430.h." What is this and how does it help you in writing programs for the microcontroller?

© The Author(s), under exclusive license to Springer Nature Switzerland AG 2023
J. Kretzschmar et al., *MSP430 Microcontroller Lab Manual*, Synthesis Lectures on Digital
Circuits & Systems, https://doi.org/10.1007/978-3-031-26643-0_2

3. In assembly code, the statement:

```
mov.b   #BIT0|BIT6, &P1OUT
```

tells the microcontroller to make pins P1.0 and P1.6 on Port 1 HIGH (puts 3.3 volts on the pins and turns them on). Alternative ways to write this command are:

```
mov.b     #0x41, &P1OUT          (Hex notation)
mov.b     #01000001b, &P1OUT     (Binary notation)
mov.b     #65, &P1OUT            (Decimal notation)
```

What are the three alternative ways of writing the following statement?

```
mov.b     #BIT1|BIT5, &P1OUT
```

2.3.2 Task B

First we walk through setting up and writing/running a simple assembly code program and then using the same steps we write/run a very simple C code program. Changing over to

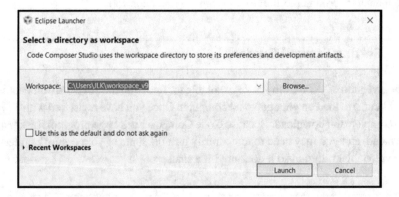

Fig. 2.1 Code Composer

Fig. 2.2 Code CCS Project

C requires a modification in Step #4 select an "Empty Project (with main.c)" instead of "Empty Assembly–only Project."

1. Upon entering into Code Composer the first screen you see is telling you where your projects are saved on your computer. This is where you go to find projects that you have previously worked on, or where new projects are stored. If you want to store your projects elsewhere (such as USB drive) select the browse button and locate a place for storage. Checking the box "Use this as the default" automatically stores projects in the workspace. Hit "Launch" and continue as shown In Fig. 2.1.
2. You now see the screen as shown in Fig. 2.2. Go to "File" button at the top select "New" and select "CCS Project." If you have a saved program you want to work with, go to "File" select "Open" and find your project in the workspace section.
3. This next screen (Fig. 2.3) allows you to name your project, select the family of micro-controller you are using, select the microcontroller chip you are using, find the USB

Fig. 2.3 Code CCS Project specifications

COM port on your computer that Code Composer wants to use, and select what type of empty project you want (assembly or C).

Selecting the "Empty Project Type" gives you a framework and has all of the setup information in place (i.e. header files). All you need to do is put in your code.

A short discussion about header files are in order. At the top of the program skeletons you see: "msp430.h" (in the assembly skeleton), and "#include <msp430>" (in the C skeleton). These are called header files and are like a dictionary that defines short names/letters to represent numbers (machine code); this allows you to make the program code you write more understandable. In the example from the Prelab, the header file "msp430.h" defines $BIT6$ as $0x40$ in hexadecimal code format. It is much easier to read

Fig. 2.4 Code CCS Project specification result

BIT6 and know what it represents than to constantly be looking up numbers. A copy of the "msp430.h" is available in the Appendices.

For "Target: " from the drop down menu select "MSP430Gxxx Family." From the next box to the right select the chip you are using "MSP430G2553." For "Connection:" click on "Identify" this searches for which USB COM port you have your LaunchPad board plugged in to. When it has found the correct USB port, you see a message box that says "Debug server session is running" now click on the "Finish" button.

For "Project Name: " give a name to your project. Just below the name you can select where you want your project to be stored. On your home computer this is probably the default location (the workspace). Next select what type of empty project you want to write code for (Assembly or C). Once all the boxes have been completed you are allowed to proceed by clicking on the "Finish" button.

The screenshot shown in Fig. 2.4 provides a completed example.

4. Figure 2.5 shows the framework for writing an assembly program. For this lab we add our code beginning at line #26. In the weeks ahead you may be writing bigger projects

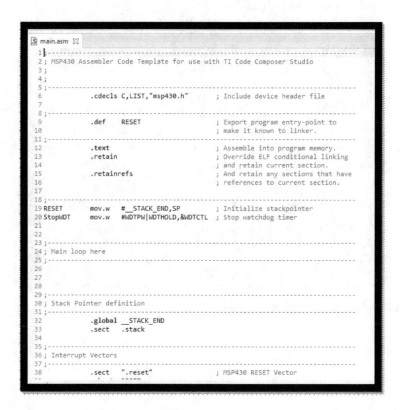

Fig. 2.5 Framework for writing an assembly program

where you may be defining a different clock speed for the CPU, or designating multiple input/output pins. The details for setting up your particular parameters can be found in the MSP430G2553 datasheet (SLAS735J), and the MSP430 Users Guide (SLAU144K). Both of these documents are available at the Texas Instruments website (www.ti.com).

5. Enter the sample assembly code as it appears in the screenshot in Fig. 2.6. This is a simple oscillator that produces a square wave at pin P1.0 on the LaunchPad.
   Note: The label "OSCILLATE" on line #27 needs to be far left (no indentation), otherwise you get an error message. All labels need to follow this format.

```
18 ;------------------------------------------------------------
19 RESET          mov.w    #__STACK_END,SP           ; Initialize s
20 StopWDT        mov.w    #WDTPW|WDTHOLD,&WDTCTL    ; Stop watchdc
21
22
23 ;------------------------------------------------------------
24 ; Main loop here
25 ;------------------------------------------------------------
26                mov.b    #BIT0, &P1DIR
27 OSCILLATE:
28                bis.b    #BIT0, &P1OUT
29                nop
30                nop
31                nop
32                bic.b    #BIT0, &P1OUT
33                nop
34                nop
35                nop
36                jmp      OSCILLATE
37
38
39 ;------------------------------------------------------------
40 ; Stack Pointer definition
41 ;------------------------------------------------------------
42                .global  __STACK_END
43                .sect    .stack
44
```

Fig. 2.6 Simple oscillator assembly program

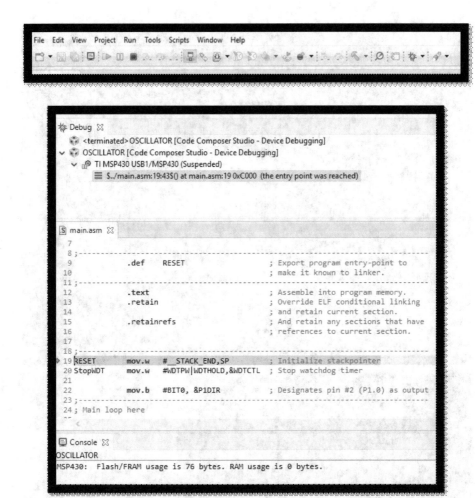

Fig. 2.7 Simple oscillator assembly program debug

6. When you have the code written, go to "File" and "Save your work." Next click on the "BUG" button (looks like a green beetle). This debugs the program and shows any errors you may have. If there are no errors a "Build Finished" briefly appears near the bottom of the screen then you see the following screen as shown in Fig. 2.7.

7. At the end of the Debug operation the program is loaded into the microcontroller. Go to the "Run" button at the top and from the drop down menu select "Resume. " As you make changes in the program follow the same sequence: (1) Save (2) Debug (3) Run (4) Resume.

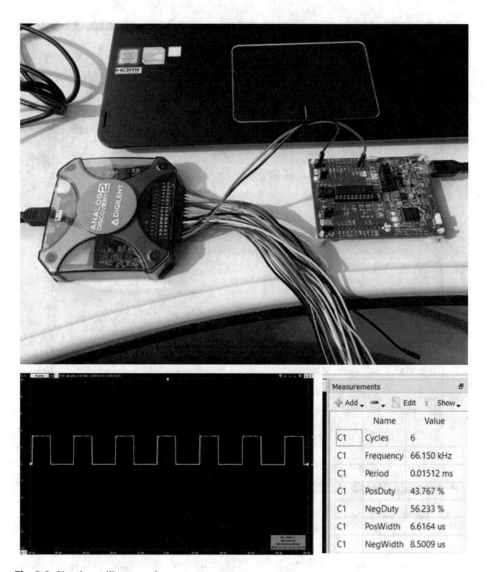

Fig. 2.8 Simple oscillator result

Your program should now be running. The green LED on P1.0 is turning on and off so fast that it only appears to you as being "ON." See the picture in Fig. 2.8 which shows it oscillating at a frequency of 66.15 KHz. Congratulations on writing your first assembly program to run the MSP430G2553 microcontroller!

As you learned in Lab #1 measurements can be made of the signal. For this signal the frequency and period are shown. Looking closely at the signal and notice that the amount of time the signal is "ON" is different from the amount of time that the signal is "OFF." This is due to the execution times needed to accomplish the instructions that were inputted into the microcontroller. You will learn more about this as you progress through the lab exercises.

Note: Programming in assembly code gives greater control for getting your microcontroller to do exactly what you want it to do, and is good for smaller programs. Higher level languages such as C, and the MSP430 specific ENERGIA (similar to Arduino coding) are nice and allow you to develop a project quickly, however you cannot really get "under the hood" and manipulate specific aspects of the microcontroller as easily as you can in assembly.

2.4   Empty Project (with main.c)

Refer back to Step #4 and select "Empty Project (with main.c)" instead of the "Empty Assembly–only Project" as we did earlier. The screen shown in Fig. 2.9 appears which is a skeleton for writing your code in the C language.

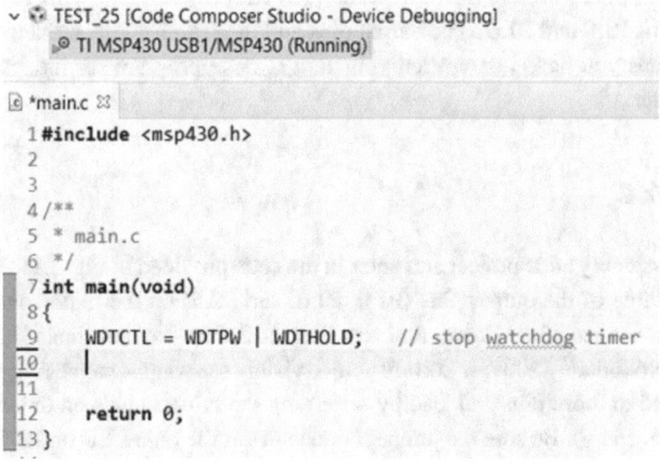

```
∨ 🔳 TEST_25 [Code Composer Studio - Device Debugging]
    🔎 TI MSP430 USB1/MSP430 (Running)

📄 *main.c ⊠
 1 #include <msp430.h>
 2
 3
 4 /**
 5 * main.c
 6 */
 7 int main(void)
 8 {
 9     WDTCTL = WDTPW | WDTHOLD;    // stop watchdog timer
10     |
11
12     return 0;
13 }
```

Fig. 2.9 Example main.c

```
main.c ✕
  1 #include <msp430.h>
  2
  3
  4 /**
  5  * main.c
  6  */
  7 int main(void)
  8 {
  9     WDTCTL = WDTPW | WDTHOLD;    // stop watchdog timer
 10
 11     P1DIR = 0x41; // Designates pins P1.0 and P1.6 as outputs
 12
 13     P1OUT = 0x41; // Turns on Red and Green LEDS
 14
 15     return 0;
 16 }
 17
```

Fig. 2.10 Example main.c code

Enter the sample C code as it appears in the screenshot provided in Fig. 2.10. Follow the same steps as in Step #8: (1) Save (2) Debug (3) Run (4) Resume. This simple program designates pins P1.0 and P1.6 to be output pins and then turns on the red and green LEDs. Congratulations you have just written your first C code program for the MSP430G2553 microcontroller.

2.5 Task C

Start a new assembly code project and enter in the code provided in Fig. 2.11. This program turns on/off three of the output pins (P1.0, P1.6, and P2.5) on the microcontroller. These three pins are connected to Green, Red and Blue LEDs on the launchpad. Each pin stays "ON" for approximately 200 ms. You are able to visibly see what is happening with the three LEDs attached to those pins and also by observing the three signals on the logic analyzer (channels 0, 5, and 6). Be sure the jumper connector for the blue LED on P2.5 is in place.

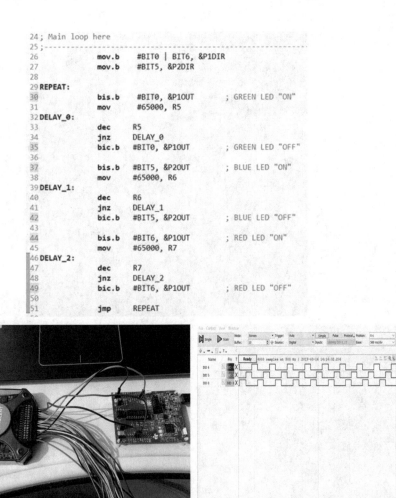

Fig. 2.11 Example main.c code and logic analyzer output

2.6   Post Lab Activities

1. In the Appendices we provide a brief introduction to Unified Modeling Language (UML)
   activity diagrams. These are UML compliant flow charts. Provide UML activity diagrams
   for programs provided in this laboratory exercise.
2. For assembly language programs provided; rewrite in C, compile, execute, and test for
   proper operation.

Lab 3: Introduction to the GP Digital I/O Pins

3

3.1 Purpose

In the first two labs you have already had some exposure to using pins on the MSP430G2553 configured as output pins. In this lab we gain experience in designating pins to be input pins as well as output pins. In addition, to the 8 pins on Port 1 (P1.0 to P1.7) you are familiar with, there is also a Port 2 with 8 pins (P2.0 to P2.7). When you review the User's Guide (SLAU144K) you see that upon power on P2.6 and P2.7 are not configured as input/output pins. To use these pins as input/output pins they need to be declared as such with P2SEL. In total there are 16 general purpose digital I/O pins that can be utilized.

Since you already have some experience with designating an output pin from the previous lab, the Discussion section below predominantly focuses on how to setup a pin as an input. In the first part of the lab a pushbutton is connected to P1.3 that is designated as an input pin. In the second part of the lab several pushbuttons are connected to pins configured as inputs. Two of these pushbuttons are used to input the "0s" and "1s" to generate a 24 bit binary number. The third pushbutton is a "Trigger" that sends the 24 bit number to a RGB LED (Red–Green–Blue). The 24 bit number defines the color and intensity of the light produced by the RGB LED, 8 bits for Red, 8 bits for Green, and 8 bits for Blue.

3.2 Discussion

It is possible to read the state of any pin (P1.0 to P1.7 and P2.0 to P2.7) of the MSP430G2553 when it is configured as an input pin. The register P1DIR allows us to select which pins we want as OUTPUT and which pins we want as INPUT. A "0" selected for a pin designates it as an INPUT, and a "1" selected for a pin designates it as an OUTPUT. Using binary notation

© The Author(s), under exclusive license to Springer Nature Switzerland AG 2023
J. Kretzschmar et al., *MSP430 Microcontroller Lab Manual*, Synthesis Lectures on Digital Circuits & Systems, https://doi.org/10.1007/978-3-031-26643-0_3

makes it easier to follow and understand what is being done as we setup and configure the I/O pins with the different registers of the I/O module.

Configuring a Pin as an INPUT:

```
mov.b #00000000b, &P1SEL   ;Selects pins P1.0 - P1.7 as GPIO pins
mov.b #01000000b, &P1DIR   ;Selects pin P1.6 as OUTPUT pin
                           ;other pins are INPUT pins
```

Because pins P1.0, P1.1, P1.2, P1.3, P1.4, P1.5, and P1.7 are selected to be INPUTs we need to enable the Pull–up/Pull–down resistors using the P1REN register, and then choose either Pull-up or Pull-down using the P1OUT register. Pull–up means that the pin (through the resistor) is internally connected to the +3.3 voltage, and Pull–down means that the pin (through the resistor) is internally connected to ground. One or the other needs to be selected or else the pin is in a "Float" status and can pick up stray signals and static causing erratic behavior. Selecting the resistor (Pull–up or Pull–down) is done through the P1OUT Register once the P1REN has been turned on. A "0" in the P1OUT Register selects the Pull–down resistor and a "1" selects the Pull–up resistor.

```
mov.b #10111111b, &P1REN  ;Enable Pull-up/Pull-down resistors
                          ;Register (P1.6 is an output)
mov.b #00010010b, &P1OUT  ;Selects Pull-up resistor for
                          ;pin P1.1 and pin P1.4
```

Special Note: Many modules of the microcontroller are multiplexed and some registers have shared functions. The "P1OUT" register is used to turn output pins on (high) or off (low), and is also used to select pull–up or pull–down resistors when a pin is designated as an input pin. Because of this, both parts need to be added together throughout the program or else the program has errors. For example say that you designate the pull–up resistor on P1.3 (mov.b #00001000b, &P1OUT) and also turn on the GREEN LED on P1.0 (bis.b #00000001b, &P1OUT) the statement needed to keep P1.3 functioning and to turn on P1.0 needs to be (bis.b #00001001b, &P1OUT).

3.3   Reading a Logic High Voltage at P1.4

If we are looking to see if a logic high voltage appears on pin P1.4, we need to ensure that a Pull–Down resistor is enabled on P1.4 so that the pin would normally be low, and not in an undefined or floating state. The code is written such that we are "testing" pin P1.4 for a situation where the pin is no longer grounded (i.e. a positive voltage has appeared on the pin). Reading a logic low voltage is the opposite, therefore, a Pull–up resistor is needed.

```
        mov.b  #00000000b, &P1SEL   ;Selects pins P1.0 -P1.7 as GPIO pins
        mov.b  #01000000b, &P1DIR   ;pin P1.6 as OUTPUT, others INPUT
        mov.b  #10111111b, &P1REN   ;Enable Pull-up/Pull-down resistors
                                    ;on INPUT pins
        mov.b  #10101111b, &P1OUT   ;Select Pull-down resistor P1.4

TEST:
        bit.b  #00010000b, &P1IN    ;Testing pin P1.4
        jnz    ON                   ;Jump if not zero-if a positive
                                    ;voltage on pin P1.4 goto ON
        jmp    TEST                 ;If still zero goto TEST
                                    ;and keep on testing

ON:                                 ;What happens when P1.4 goes High
```

3.4   Reading a Logic Low Voltage at P1.4

```
        mov.b  #00000000b, &P1SEL   ;Select pin as GPIO and clear P1 Port
        mov.b  #01000000b, &P1DIR   ;Select pin P1.6 as OUTPUT,
                                    ;other pins are INPUT
        mov.b  #10111111b, &P1REN   ;Enables Pull-up/Pull-down resistors
        mov.b  #10111111b, &P1OUT   ;Selects the Pull-up resistor on P1.4

TEST:
        bit.b  #00010000b, &P1IN    ;Testing pin P1.4
        jz     ON                   ;Jump if zero-if a logic low voltage
                                    ;on pin P1.4 goto ON
        jmp    TEST                 ;If still not zero keep on testing

ON:                                 ;What happens when P1.4 goes low
```

3.5   Switch Bounce

The microcontroller CPU operates very fast (for a 16 MHz clock speed, one cycle time is 0.0625 μs). A "bit.b" instruction takes two cycle times or 0.125 μs. A switch may stay pushed for well over a millisecond and when released goes through a "transition" period where the contacts go from connected to not connected. This transition is called "switch bounce" and may appear to the microcontroller as if the switch has been repeatedly turned on and off for a very short duration of time." Meanwhile the CPU has executed the program statement several times–possibly detecting the unwanted switch bounce states. The switch has caused the problem because during the "transition period" the program does not know how to interpret what is happening. There are several ways to manage switch bounce. A software approach is to place a delay in the program after a switch is read, and before the next line of code is executed (60 ms generally works well).

3.6   PreLab

3.6.1   Task A

1. Read through the documentation for this lab so you are well prepared for the other Tasks.
2. Connect a pushbutton to your MSP430G2553 microcontroller on pin P2.2 as depicted in Fig. 3.1. You are tasked with monitoring the status of this pin and whether the pushbutton has been pushed. Using the assembly code given above in the Discussion write the code for monitoring pin P2.2. Make all the pins of Port 2 input pins. **Hint**: Use the "Reading a Logic Low Voltage" example.

3.6.2   Task B

This task is a simple C program (DL_SWITCH_TEST) to demonstrate the function of reading the switch on pin P1.3 of the LaunchPad. If the button is not pushed then P1.0 is turned on and the GREEN LED lights up. If the button is pushed then P1.0 goes off and P1.6 is turned on and the RED LED lights up. The code listing is provided below and a detailed explanation of what is going in the program is in the comment section. In Code Composer select an "Empty Project (with main.c)," type in the code, and run the program.

```
//************************************************************************
// Program Name: SWITCH_TEST
// C Code using Texas Instruments Code Composer
// Parts:
// -- Texas Instruments MSP-EXP430G2ET Launchpad with MSP430G2553
//
// Documentation:
// -- Texas Instruments SLAU144K Users Guide (ti.com)
// -- Texas Instruments SLAS735J MSP430G2553 Datasheet (ti.com)
//
// This program sets up P1.0 and P1.6 as output pins. The GREEN LED on the
// LaunchPad is connected to P1.0 and the RED LED is connected to P1.6.
// P1.3 is setup as an input pin and the pushbutton on the side of the
// LaunchPad is connected to P1.3 (one pole to P1.3 and the other pole to
// Ground), therefore, a pullup resistor is needed. This program runs a loop
// that continually checks to see if the switch on P1.3 has been pushed
// which would ground pin P1.3 (place a "0" on it). If not pushed the
// GREEN LED (P1.0) stays on, and if pushed the GREEN LED (P1.0) is turned
// off and the RED LED (P1.6) is turned on.
//
// NOTE:
```

Fig. 3.1 Code Composer

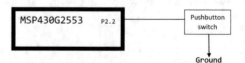

```
// Many pins in the microcontroller are multiplexed and can be selected to perform
// shared functions. The "P1OUT" register is used to send signals out on Port 1 pins,
// and is also used to select pull-up or pull-down resistors when a pin is designated
// as an input pin. Because of this dual function, both parts (pins selected to send
// signals out and pins with pullup/pulldown resistors selected) need to be combined
// throughout the program or else the program will have errors. For example, when
// turning on the GREEN LED (P1OUT = 0b00000001) (Hex 0x01), P1OUT previously had the
// pullup resistor selected for P1.3 (P1OUT = 0b00001000) (Hex 0x08), therefore to
// preserve the pullup resistor on P1.3 the statement should be (P1OUT = 0b00001001)
// (Hex 0x09). Both Port 1 and Port 2 in the MSP430G2553 function in this manner.
//
// J. Kretzschmar

#include <msp430.h>

int main(void)
{
    WDTCTL = WDTPW | WDTHOLD; // stop watchdog timer
    P1SEL = 0b00000000;    // Selects all Port 1 pins as an I/O pins (Hex 0x00)
    P1DIR = 0b01000001;    // P1.0 and P1.6 are output pins (Hex 0x41)
                           // The switch on the Launchpad is connected to P1.3 and
                           // Ground. P1.1  P1.2  P1.4  P1.5  P1.7 are also input pins
    P1REN = 0b00001000;    // Engages the pull-up/down resistors for P1.3 (Hex 0x08)
    P1OUT = 0b00001000;    // Selects the pull-up resistor on P1.3 (Hex 0x08)
                           // See note about multiplexing pins of Port 1

    for(;;)                // This is an infinite loop
    {
      if (P1IN & 0b00001000)  // If the condition is TRUE (P1IN = 1 & BIT3 =1)
                           // then the switch has not been pushed (a "1" is on
                           // P1.3), therefore, the next statement is executed
                           // (GREEN LED is turned on)
          P1OUT = 0b00001001; // GREEN LED turned on (Hex 0x09)
                           // Pullup resistor on P1.3 preserved
                           // See note about multiplexing pins of Port 1

      else                 // If the switch is pushed then P1.3 is grounded
                           // (a "0" is on P1.3) which makes the "if" statement
                           // false and the program executes the "else" statement
                           // which turns off the GREEN LED and turns on the RED
                           // LED.
          P1OUT = 0b01001000; // RED LED turned on (Hex 0x48)
                           // Pullup resistor on P1.3 preserved
                           // See note about multiplexing pins of Port 1
    }
}

//*****************************************************************************
```

3.6.3   Task C

This task is an assembly program (DL_RGB_10) demonstrating the use of multiple input pins and multiple output pins. Two pushbutton switches (P1.2 and P1.3) are used to input a string of 24 "0s" and "1s" (a 24 bit number) that is stored in two 16–bit registers. Once 24 bits have been inputted the GREEN LED (P1.0) comes on to inform you that 24 bits have been entered. The third pushbutton switch (P1.7 the trigger) is pushed and the microcontroller now creates the pulses that represent the 24 bit number and sends them to a Red–Green–Blue (WS2812B) LED via pin P1.6. The 24 bit number defines the color that is produced by the RGB LED. The first eight bits define the amount of Green, the second eight bits define the amount of Red, and the third eight bits define the amount of Blue. Mixing different amounts of each defines the color produced.

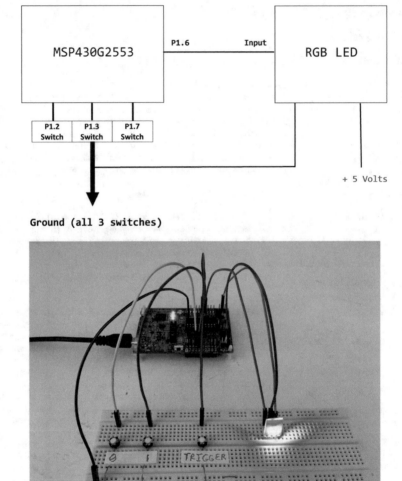

Fig. 3.2 Circuit configuration

The code listing is provided below and a detailed explanation of what is going on in the program is in the comment section. In Code Composer select an "Empty Project Assembly--only Project," type in the code, and run the program. The circuit configuration is provided in Fig. 3.2.

```
;-------------------------------------------------------------------------------
; Program Name: INPUT_RGB_LED
; Assembly Code using Texas Instruments Code Composer
; Parts:
; -- Texas Instruments MSP-EXP430G2ET Launchpad with MSP430G2553
; -- Worldsemi WS2812B RGB LED
; -- 3 Tactile switches
; Documentation:
; -- Texas Instruments SLAU144K Users Guide (ti.com)
; -- Texas Instruments SLAS735J MSP430G2553 Datasheet (ti.com)
; -- Worldsemi WS2812B datasheet (world-semi.com)
; -- Paper "Interfacing the Worldsemi Electronics WS2812B Addressable RGB
;    Light Emitting Diode with a Texas Instruments MSP430 Microcontroller"
;    (Appendix)
; Connections:
; -- Tactile Switch (Send Color): one pole to P1.7 and one pole to Ground
; -- Tactile Switch ("0"): one pole to P1.2 and one pole to Ground
; -- Tactile Switch ("1"): one pole to P1.3 and one pole to Ground
; -- WS2812B RGB LED: +5 on Launchpad
; -- WS2812B RGB LED: Ground on Launchpad
; -- WS2812B RGB LED: Data Line input connected to P1.6
;
; This program inputs a 24-Bit binary number which defines a specific color into
; the WS2812B RGB LED through pushbutton switches on pins P1.2 (inputs a "0")
; and P1.3 (inputs a "1"). When 24 bits have been inputted the GREEN LED on pin
; P1.0 is turned on to indicate that 24 bits have been entered. When the "Send
; Color" pushbutton switch on P1.7 is pushed the 24-Bit number is sent out to
; the RGB (RED-GREEN-BLUE) LED on pin P1.6. The color produced is dependent upon
; the intensity of the 3 colors. The intensity is defined by the value (0-255)
; given to each color. The first 8 bits define the intensity of GREEN. The
; second 8 bits define the intensity of RED. The third 8 bits define the
; intensity of BLUE. For example to produce a purple color the following
; intensity values would be entered: (GREEN = 0  RED = 255  BLUE = 255). This
; is 8 pushes of the "0" button, 8 pushes of the "1" button, and 8 pushes of
; the "1" button for a total of 24 button pushes. Enter a new 24 bit number to
; produce a new color.
;
; J. Kretzschmar


;-------------------------------------------------------------------------------
          .cdecls C,LIST,"msp430.h"      ; Include device header file
          .def     RESET
          .text
          .retain
          .retainrefs
;-------------------------------------------------------------------------------
RESET     mov.w    #__STACK_END,SP       ; Initialize stackpointer
StopWDT   mov.w    #WDTPW|WDTHOLD,&WDTCTL ; Stop watchdog timer
;-------------------------------------------------------------------------------
;----- ESTABLISHING CLOCK SPEED AND INPUT AND OUTPUT PINS ----------------------
;-------------------------------------------------------------------------------
NEW_COLOR:
```

```
      mov.b   &CALBC1_1MHZ,&BCSCTL1   ; 1 MHz clock
      mov.b   &CALDCO_1MHZ,&DCOCTL    ; 1 MHz MCLK

      mov.b   #01110001b, &P1DIR      ; P1.0/P1.4/P1.5/P1.6 are output pins
                                      ; P1.1/P1.2/P1.3/P1.7 are input pins
      mov.b   #10001100b, &P1REN      ; Enables Pullup/Pulldown resistors on
                                      ; input pins P1.2/P1.3/P1.7
      mov.b   #10001100b, &P1OUT      ; Enables Pullup resistors on
                                      ; input pins P1.2/P1.3/P1.7

      mov     #0, R6                  ; Clears R6  (24 button push counter)
      mov     #0, R10                 ; Clears R10 (RGB data storage register)
      mov     #0, R11                 ; Clears R11 (RGB data storage register)
      mov     #0, R12                 ; Clears R12 (24 bits delivered counter)
;-------------------------------------------------------------------------------
;----- CHECKING SWITCHES FOR PUSHES --------------------------------------------
;-------------------------------------------------------------------------------
TEST:                                 ; Testing switches on P1.2 and P1.3

      cmp     #24, R6                 ; Has there been 24 pushes?
      jeq     ENTERED_24_BITS         ; If yes goto ENTERED_24_BITS

      bit.b   #00000100b, &P1IN       ; Was P1.2 switch pushed? (the "0" switch)
      jz      INPUT_A_ZERO            ; If yes goto INPUT_A_ZERO
      bit.b   #00001000b, &P1IN       ; Was P1.3 switch pushed? (the "1" switch)
      jz      INPUT_A_ONE             ; If yes goto INPUT_A_ONE

      jmp     TEST                    ; If no keep testing P1.2 and P1.3
;-------------------------------------------------------------------------------
;----- INPUT A "0" INTO REGISTERS HOLDING THE 24 BIT NUMBER (R10 and R11) ------
;-------------------------------------------------------------------------------
INPUT_A_ZERO:
      mov     #65535, R5              ; Delay so pushbutton is not read too fast
DELAY_1:                              ; Delay routine
      dec     R5                      ; Delay routine
      jnz     DELAY_1                 ; Delay routine

      add     R10, R10                ; A "0" is inputted into R10
      addc    R11, R11                ; When R10 is full overflow goes into R11
      add     #1, R6                  ; Add 1 to the 24 button push counter
      jmp     TEST                    ; Goto TEST to continue testing
;-------------------------------------------------------------------------------
;----- INPUT A "1" INTO REGISTERS HOLDING THE 24 BIT NUMBER (R10 and R11) ------
;-------------------------------------------------------------------------------
INPUT_A_ONE:
      mov     #65535, R5              ; Delay so pushbutton is not read too fast
DELAY_2:                              ; Delay routine
      dec     R5                      ; Delay routine
      jnz     DELAY_2                 ; Delay routine

      add     R10, R10                ; A "0" is inputted into R10
      addc    R11, R11                ; When R10 is full overflow goes into R11
      bis     #1, R10                 ; A "1" is now placed into R10
      add     #1, R6                  ; Add 1 to the 24 button push counter
      jmp     TEST                    ; Goto TEST to continue testing
;-------------------------------------------------------------------------------
;----- 24 NUMBERS HAVE BEEN ENTERED AND THE GREEN LED ON P1.0 TURNS ON --------
;-------------------------------------------------------------------------------
ENTERED_24_BITS:
```

```
      bis.b    #00000001b, &P1OUT      ; GREEN LED (P1.0) turns on when R6 = 24
;----------------------------------------------------------------------------
;----- PREPARING THE 24 BIT NUMBER FOR LOADING INTO THE WS2812B LED -----------
;----------------------------------------------------------------------------
;----- Registers R11 and R10 now hold the 24 Bit number. Since the -----------
;----- upper 8 Bits of R11 do not hold any of the inputted 24 Bits, 8 "0"s ----
;----- need to be loaded into the lower 8 Bits of R10 so that the 24 Bit ------
;----- number is ready to be loaded into the RGB LED. The 24 Bit number is ----
;----- pushed out of R11 through the Carry Bit which test for the Bit being ---
;----- a "1" or a "0". The WS2812B RGB LED needs specific timing for the ------
;----- "1"s and "0"s and this timing is accomplished by the MAKE_A ZERO and ---
;----- MAKE_A_ONE sections.
;----------------------------------------------------------------------------
      mov      #8, R8                  ; The number 8 is loaded into R8
EIGHT_ZEROS:                           ; This routine inputs 8 more "0s" into R10
      add      R10, R10                ; this pushes the 24 bits already inputted
      addc     R11, R11                ; so that R11 is completely filled up with
      dec      R8                      ; RGB color data to be sent
      jnz      EIGHT_ZEROS

      jmp      SEND_COLOR_DATA         ; Goto SEND_COLOR_DATA
;----------------------------------------------------------------------------
;----- CHANGES CLOCK SPEED SO TIMING IS CORRECT FOR DATA ENTRY ---------------
;----- TEST P1.7 SWITCH FOR SENDING DATA TO WS2812B LED ----------------------
;----------------------------------------------------------------------------
SEND_COLOR_DATA:
      mov.b    &CALBC1_16MHZ,&BCSCTL1  ; Change CPU clock to 16 MHz
      mov.b    &CALDCO_16MHZ,&DCOCTL   ; Change CPU clock to 16 MHz
      bit.b    #10000000b, &P1IN       ; Was P1.7 switch pushed?
      jz       READ_A_BIT              ; If yes goto READ_A_BIT
      jmp      SEND_COLOR_DATA         ; If no keep on testing P1.7
;----------------------------------------------------------------------------
;----- ROUTINE THAT SENDS 24 BITS TO THE WS2812B RGB LED ---------------------
;----------------------------------------------------------------------------
READ_A_BIT:
      cmp      #24, R12                ; Have 24 bits been delivered?
      jeq      NEW_COLOR               ; If yes goto the NEW_COLOR
      add      R10, R10                ; This pushes 1 Bit out the MSB end of R11
      addc     R11, R11                ; into the "Carry Bit"
      jc       MAKE_A_ONE              ; Is the "Carry Bit" set (have a "1" in it)?
      jmp      MAKE_A_ZERO             ; If yes goto MAKE_A_ONE
                                       ; If no goto MAKE_A_ZERO
;----------------------------------------------------------------------------
;----- MAKES THE "0"s WITH THE REQUIRED TIMING PARAMETERS --------------------
;----- SEE THE WS2812B DATASHEET FOR DETAILS --------------------------------
;----- MEASURED ON OSCILLOSCOPE ---------------------------------------------
;----------------------------------------------------------------------------
MAKE_A_ZERO:
      bis.b    #01000000b, &P1OUT      ; Approximately 350ns HIGH
      nop
      nop
      bic.b    #01000000b, &P1OUT      ; Approximately 900ns LOW
      nop
      nop
      nop
      nop
      nop
      nop
      add      #1, R12                 ; Add 1 to the counter
```

```
    jmp     READ_A_BIT              ; Go read another bit
;-----------------------------------------------------------------------
;----- MAKES THE "1"s WITH THE REQUIRED TIMING PARAMETERS --------------
;----- SEE THE WS2812B DATASHEET FOR DETAILS --------------------------
;----- MEASURED ON OSCILLOSCOPE ---------------------------------------
;-----------------------------------------------------------------------
MAKE_A_ONE:
    bis.b   #01000000b, &P1OUT      ; Approximately 900ns HIGH
    nop
    nop
    nop
    nop
    nop
    nop
    nop
    nop
    nop
    nop
    bic.b   #01000000b, &P1OUT      ; Approximately 350ns LOW
    nop
    nop
    add     #1, R12                 ; Add 1 to the counter
    jmp     READ_A_BIT              ; Go read another bit
;-----------------------------------------------------------------------
; Stack Pointer definition
;-----------------------------------------------------------------------
            .global __STACK_END
            .sect   .stack
;-----------------------------------------------------------------------
; Interrupt Vectors
;-----------------------------------------------------------------------
            .sect   ".reset"                ; MSP430 RESET Vector
            .short  RESET

;-----------------------------------------------------------------------
```

3.7   Post Lab Activities

1. In the Appendices we provide a brief introduction to Unified Modeling Language (UML) activity diagrams. These are UML compliant flow charts. Provide UML activity diagrams for programs provided in this laboratory exercise.
2. For assembly language programs provided; rewrite in C, compile, execute, and test for proper operation.

Lab 4: Exploring the MSP430G2553 Basic Clock Module

4

4.1 Purpose

Every computer system (microcontrollers are small computers) contains an internal clock that regulates how quickly instructions can be executed. The MSP430G2553 microcontroller CPU uses clock signals to fetch and execute instructions and also in the operation of peripheral modules (such as the timer module). The Basic Clock Module of the MSP430 has three available clock signals:

- ACLK: This is the Auxiliary clock and is used in peripheral modules.
- MCLK: This is the Master clock and is used by the CPU.
- SMCLK: This is the Sub–main clock and is used in the peripheral modules.

For the MCLK and the SMCLK the clock signal can be generated from the internal DCO (Digitally Controlled Oscillator), the internal VLO (Very Low power and Low frequency Oscillator), or an external crystal. The ACLK clock signal can be generated from the VLO or a crystal. In this lab we investigate the DCO and the VLO clock sources and how the speed of these clock sources can be changed to fit a particular requirement for a project.

It is possible to use a crystal oscillator source to input a clock signal into the microcontroller, however, this lab only investigates sources internally available on the microcontroller. For detailed information about these clock sources see the MSP430G2553 Datasheet (SLAS735J) and the MSP430 Users Guide (SLAU144K).

© The Author(s), under exclusive license to Springer Nature Switzerland AG 2023   29
J. Kretzschmar et al., *MSP430 Microcontroller Lab Manual*, Synthesis Lectures on Digital Circuits & Systems, https://doi.org/10.1007/978-3-031-26643-0_4

4.2   Discussion

After power on, the MSP430G2553 by default sources the MCLK and the SMCLK from the DCO at a frequency of approximately 1.1 MHz. If needed, the clock frequency can be changed from the default setting by making adjustments to the appropriate registers. There are four pre–determined calibrated frequencies stored internally that can be used (1, 8, 12 and 16 MHz), and it is possible to select a unique desired frequency by adjusting the DCO clock via the control registers. The DCO clock is susceptible to temperature variations because the frequency of the oscillator is determined by a RC (resistor/capacitor) network. If higher precision is necessary the clock system can be sourced externally from a crystal oscillator.

There are four Basic Clock Module registers that can be adjusted to setup the clock frequency that is needed and each register controls different parameters of the clock system. It can be daunting at first, however, for most situations you only need to engage a couple of the parameters in each register to setup what is needed for a project. The full details for each register are available in the User's Guide. The registers include:

- DCOCTL: Digitally Controlled Oscillator (DCO) Control Register
- BCSCTL1: Basic Clock System Control Register 1
- BCSCTL2: Basic Clock System Control Register 2
- BCSCTL3: Basic Clock System Control Register 3

In this lab you use two programs (C code and Assembly code) to investigate the clock system. Each program is well commented so you can understand what is happening. In both of the programs most of the instructions are "commented out," and as you "comment in" these instructions the clock frequency is adjusted. The program itself is a simple loop that is setup to function as an oscillator so that you can visualize (on an oscilloscope connected to pin P1.0) the effect of the changes you make to the clock system.

4.3   PreLab

4.3.1   Task A

1. Read through the documentation for this lab so you are well prepared for the other Tasks.
2. Review the diagram in Fig. 4.1. The clock system default frequency of the MSP430G2553 is approximately 1.1 MHz and has (saved in memory) DCO settings of RSELx = 7 (BCSCTL1 Bits 3–0) and DCOx = 3 (DCOCTL Bits 7–5).
   The Assembly code statements to set these parameters are as follows:

```
mov.b    #00000111b, &BCSCTL1
mov.b    #01100000b, &DCOCTL
```

Fig. 4.1 Default setting and approximate clock frequency

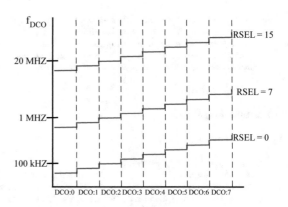

3. Write the Assembly code statements to set the DCO parameters for a clock frequency of approximately 100 KHz. It is difficult to discern exactly from this chart what the exact DCO setting should be. Please approximate this value.

4.4 Task B

For this task load the C program (DL_CLOCK_TEST) and setup an oscilloscope attached to pin P1.0. Run the program and observe the changes in the frequency of the waveform as you change the clock speeds from the default setting (no additional code) to the predetermined calibrated settings of 1, 8, 12 and 16 MHz. You should change the DCO frequency by uncommenting the two lines of code associated with each frequency.

```
//-------------------------------------------------------------------------
// Program Name: CLOCK_TEST
// C Code using Texas Instruments Code Composer
// Parts:
// -- Texas Instruments MSP-EXP430G2ET Launchpad with MSP430G2553
// -- Oscilloscope
//
// Documentation:
// -- Texas Instruments SLAU144K Users Guide (ti.com)
// -- Texas Instruments SLAS735J MSP430G2553 Datasheet (ti.com)
//
// Connections:
//
// MSP430G2553          Oscilloscope
//      P1.0  ---------->  Input
//
// This program is a simple oscillator for exploring the clock
// system of the MSP430G2553. Connect an oscilloscope to pin
// P1.0 and view the waveform as you make changes to the speed
// of the clock. Comment "IN" the Calibrated Clock statements
// one at a time and view on an oscilloscope. The 1 MHZ clock
```

```
// is already commented "IN" and will have to be commented "OUT"
// as the 8 MHZ, 12 MHZ, and 16 MHZ clock speeds are investigated.
// The square wave on the oscilloscope does not represent the
// speed of the clock, it only shows how the CPU clock affects
// the speed of operations and is a relative indicator only.
//
// J. Kretzschmar

#include <msp430.h>

int main(void)
{
    WDTCTL = WDTPW | WDTHOLD;  // stop watchdog timer
    P1SEL = 0x00;             // Selects Port 1 as an I/O port
    P1DIR = 0x01;             // P1.0 is an output pin
    long int i;  // Declare "i" as a variable (up to 2147483647)
    long int j;  // Declare "j" as a variable (up to 2147483647)

    BCSCTL1 = CALBC1_1MHZ;  // 1 MHZ Calibrated Clock
    DCOCTL  = CALDCO_1MHZ;  // 1 MHZ

    //BCSCTL1 = CALBC1_8MHZ;  // 8 MHZ Calibrated Clock
    //DCOCTL  = CALDCO_8MHZ;  // 8 MHZ

    //BCSCTL1 = CALBC1_12MHZ; // 12 MHZ Calibrated Clock
    //DCOCTL  = CALDCO_12MHZ; // 12 MHZ

    //BCSCTL1 = CALBC1_16MHZ; // 16 MHZ Calibrated Clock
    //DCOCTL  = CALDCO_16MHZ; // 16 MHZ

    while(1)                  // While TRUE keep looping
    {
        i = 0;               // Make i = 0
        j = 0;               // Make j = 0
        P1OUT = 0x01;        // Turn on RED LED
        while(i < 200000)    // While "i" < 200000 keep LED on
        {
            i++;             // Add 1 to "i"
        }

        P1OUT = 0x00;        // "i" hit 200000 so turn RED LED off
        while(j < 200000)    // While "j" < 200000 keep LED off
        {
            j++;             // Add 1 to "j"
        }
    }
    return 0;
}

//-----------------------------------------------------------------------
```

The Figures shown in 4.2 show what to expect when you change the clock speeds.

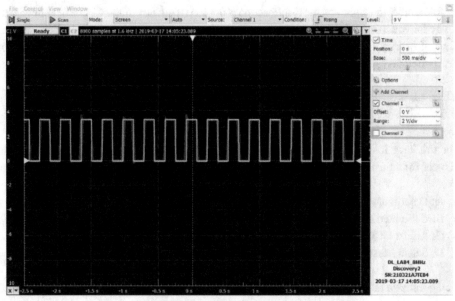

(a) 8 MHz calibrated clock.

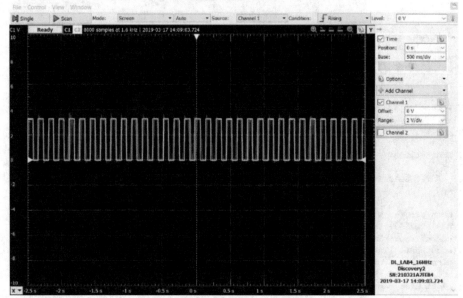

(b) 16 MHz calibrated clock.

Fig. 4.2 Expected waveforms

Remember the signal on the oscilloscope is not the frequency of the CPU clock, however, a way for you to visualize what happens when you change the CPU clock frequency. Several lines of instruction code are executed to produce the signal on the oscilloscope and each instruction takes several clock cycles (clock ticks of the CPU clock).

4.5   Task C

For this task load the Assembly program (DL_CLOCK_SELECTION). There are four choices for adjusting the speed of the clock system "commented out" in the program listing:

- A: Default clock (nothing to change here)
- B: Calibrated clock frequencies
- C: Adjust DCO clock frequency
- D: Select Master Clock (MCLK) source (DCO or VLOCLK)

Go through the different choices and observe what happens with the oscilloscope wave-form.

```
;-------------------------------------------------------------------------
; Program Name: CLOCK_SELECTION
; Assembly Code using Texas Instruments Code Composer
; Parts:
; -- Texas Instruments MSP-EXP430G2ET Launchpad with MSP430G2553
; -- Oscilloscope
;
; Documentation:
; -- Texas Instruments SLAU144K Users Guide (ti.com)
; -- Texas Instruments SLAS735J MSP430G2553 Datasheet (ti.com)
;
; Connections:
;
;    MSP430G2553          Oscilloscope
;        P2.1  ---------->  Input
;
; This program will allow for experimentation with clock sources internal to the
; MSP430G2553. The default clock (the DCO oscillator) for the CPU is internally
; set at approximately 1.1 MHz and the microcontroller will function without any
; additional lines of code. To select another clock source other than the default
; clock, lines of code need to be added to the program. There are 4 choices below
; and each has the code to engage the clock source under test. To engage a
; particular clock source "uncomment" the appropriate lines of code. The TEST
; OSCILLATOR is a means to visualize (with oscilloscope on pin P1.0) the change
; in the speed of execution of code. The square wave on the oscilloscope does
; not represent the speed of the clock, it only shows how the CPU clock affects
; the speed of operations and is a relative indicator only.
;
; J. Kretzschmar
;
;-------------------------------------------------------------------------
            .cdecls C,LIST,"msp430.h"        ; Include device header file
```

```
                .def    RESET
                .text
                .retain
                .retainrefs
;--------------------------------------------------------------------------------
RESET       mov.w   #__STACK_END,SP         ; Initialize stackpointer
StopWDT     mov.w   #WDTPW|WDTHOLD,&WDTCTL  ; Stop watchdog timer
;********************************************************************************
; CHOICE (A)                       DEFAULT CLOCK
;           Do not have to add any lines of code
;           The DCO clock is preset at a frequency of approx 1.1 MHz
;           (TEST OSCILLATOR Approximate Frequency = 66 KHz)
;********************************************************************************
;********************************************************************************
; CHOICE (B)                 CALIBRATED CLOCK FREQUENCIES
;             (Calibration values stored internally in the MSP430G2553)
;
        ;mov.b   &CALBC1_1MHZ, &BCSCTL1 ; (TEST OSCILLATOR Frequency = 62.5 KHz)
        ;mov.b   &CALDCO_1MHZ, &DCOCTL

        ;mov.b   &CALBC1_8MHZ, &BCSCTL1 ; (TEST OSCILLATOR Frequency = 490.1 KHz)
        ;mov.b   &CALDCO_8MHZ, &DCOCTL

        ;mov.b   &CALBC1_12MHZ, &BCSCTL1 ; (TEST OSCILLATOR Frequency = 757.5 KHz)
        ;mov.b   &CALDCO_12MHZ, &DCOCTL

        ;mov.b   &CALBC1_16MHZ, &BCSCTL1 ; (TEST OSCILLATOR Frequency = 990.0 KHz)
        ;mov.b   &CALDCO_16MHZ, &DCOCTL
;********************************************************************************
;********************************************************************************
; CHOICE (C)             ADJUST THE  DCO CLOCK FREQUENCY
;                          (see SLAU144K for details)
;
;               Select RSEL (Bits 0-3) for &BCSCTL1
;               Select DCO  (Bits 5-7) for &DCOCTL
;               Binary format used here for clarity (need "b" at end of number)

        ;mov.b   #00000111b, &BCSCTL1 ; Default clock setting (RSEL=7)
        ;mov.b   #01100000b, &DCOCTL  ; Default clock setting (DCO=3)
                                     ; For the block of Bits (5,6,and 7) which
                                     ; is labeled as "DCO" within the DCOCTL
                                     ; register "DCO" does equal "3". For the
                                     ; DCOCTL register it equals "96"
;********************************************************************************
;********************************************************************************
; CHOICE (D)      SELECT MASTER CLOCK (MCLK) SOURCE (DCO or VLOCLK)
;
;               Basic Clock System Control Register 2 (BCSCTL2)
;                      (see SLAU144K for details)
; BCSCTL2 is used for selecting the source clock for the MCLK (DCO or VLOCLK)
; BCSCTL2 is also used for dividing the MCLK (MCLK/1  MCLK/2  MCLK/4  MCLK/8)
;
;               Select the DCO as the clock source and divisor

        ;mov.b #00000000b, &BCSCTL2 ; DCO selected as clock: MCLK/1
        ;mov.b #00010000b, &BCSCTL2 ; DCO selected as clock: MCLK/2
        ;mov.b #00100000b, &BCSCTL2 ; DCO selected as clock: MCLK/4
        ;mov.b #00110000b, &BCSCTL2 ; DCO selected as clock: MCLK/8
```

```
;         To Select the VLOCLK as the clock source (approx 12.0 KHz)
;         Must select appropriate bits in BCSCTL1 and BCSCTL3 to engage the VLOCLK
;                   (All 3 lines of code need to be uncommented)
;
        ;mov.b    #00000000b, &BCSCTL1 ; Bit 6 must be "0" (XTS)
        ;mov.b    #00100000b, &BCSCTL3 ; Bits 5-4 must be "10" (LFXT1S)
        ;mov.b    #11000000b, &BCSCTL2 ; VLOCLK selected as clock: MCLK/1

;           Uncomment lines below to replace &BCSCTL2 and divide MCLK

          ;mov.b    #11010000b, &BCSCTL2 ; VLOCLK selected as clock: MCLK/2
          ;mov.b    #11100000b, &BCSCTL2 ; VLOCLK selected as clock: MCLK/4
          ;mov.b    #11110000b, &BCSCTL2 ; VLOCLK selected as clock: MCLK/8
;*********************************************************************************
;
;                              TEST OSCILLATOR
;
;*********************************************************************************

          mov.b    #BIT0, &P1DIR
OSCILLATE:
          bis.b    #BIT0, &P1OUT
          nop
          nop
          nop
          bic.b    #BIT0, &P1OUT
          nop
          nop
          nop
          jmp      OSCILLATE
;-------------------------------------------------------------------------------
; Stack Pointer definition
;-------------------------------------------------------------------------------
          .global  __STACK_END
          .sect    .stack
;-------------------------------------------------------------------------------
; Interrupt Vectors
;-------------------------------------------------------------------------------
          .sect    ".reset"                    ; MSP430 RESET Vector
          .short   RESET
;-------------------------------------------------------------------------------
```

4.6   Task D

Select the VLOCLK as the clock source (approximately 12 KHz) for the MCLK and divide the MCLK by 4 (MCLK/4). What is the approximate frequency of the Master Clock (MCLK)? What is the frequency of the waveform on the oscilloscope? Why are the two frequencies different?

4.7   Post Lab Activities

1. In the Appendices we provide a brief introduction to Unified Modeling Language (UML) activity diagrams. These are UML compliant flow charts. Provide UML activity diagrams for programs provided in this laboratory exercise.
2. For assembly language programs provided; rewrite in C, compile, execute, and test for proper operation.

Lab 5: Interfacing a 16x2 LCD to the MSP430G2553

5.1 Purpose

In this lab you interface a 16x2 LCD display with the MSP430G2553 microcontroller. This display is used in subsequent labs to output data from the microcontroller. There are many varieties of 16x2 LCD displays and in this lab we use one that accepts data input in parallel format (versus serial format) and has a backlight which makes it much easier to see (Fig. 5.1).

5.2 Discussion

Built into the 16x2 LCD display is a controller that generates the alpha numeric characters when it is presented with an ASCII value. The most common controllers are the HD44780, and the newer controller the ST7066U.

Both of these controllers accept the same commands. Deciphering the datasheets may be challenging, however, in the Appendices we provide a paper titled "Controlling a 16x2 LCD Display with the MSP430G2553 Microcontroller Using Assembly Code." The Appendix explains the commands to control the display. In this lab we use a 5.0 volt 16x2 LCD display with the supply voltage for the backlight on pin #15 (+5.0 volts) and pin #16 (Ground) of the display module. The appendix represents binary numbers in the format of #0b00000000, for Code Composer, the format needs to be #00000000b. Also the paper was written for the eight data lines to be connected to Port 1 pins on the microcontroller. In this lab Port 2 pins are used so follow the guide below.

© The Author(s), under exclusive license to Springer Nature Switzerland AG 2023

J. Kretzschmar et al., *MSP430 Microcontroller Lab Manual*, Synthesis Lectures on Digital Circuits & Systems, https://doi.org/10.1007/978-3-031-26643-0_5

Fig. 5.1 16x2 LCD display with the MSP430G2553 microcontroller

5.3   Prelab

5.3.1   Task A

1. Read through the documentation for this lab so you are well prepared for the other Tasks.
2. To complete this Task you have to read the paper mentioned above and review the code listing provided below.
3. In the lab file provided below, go to "DATA INPUT SEQUENCE." Write the code that displays your first name in capital letters. You have to look at an ASCII Table.
4. What would you have to do to shift what is displayed three spaces to the left? **Hint:** Look at the section in the paper "COMMAND: DDRAM Address" and find out where the "MSP430" is displayed now). For your answer give the line of code that needs to be changed, and how you would change it.

```
;------------------------------------------------------------------------
; Program Name: LCD_PORT_2
; Assembly Code using Texas Instruments Code Composer
; Parts:
; -- Texas Instruments MSP-EXP430G2ET Launchpad with MSP430G2553
; -- Newhaven NHD-0216K1Z 16x2 LCD Display (or similar display)
; -- 10K Potentiometer (for contrast control)
; Documentation:
; -- Texas Instruments SLAU144K Users Guide (ti.com)
```

```
; -- Texas Instruments SLAS735J MSP430G2553 Datasheet (ti.com)
; -- Newhaven NHD-0216K1Z Datasheet (newhavendisplay.com)
; -- Paper "Controlling a 16x2 LCD Display with the MSP430G2553 Microcontroller
;    Using Assembly Code" (Appendix)
;
; This program uses a MSP430G2553 microcontroller to drive a 16x2 LCD display
; and prints "MSP430" on the display for demonstration purposes. LCD Display
; delays are longer than noted in the datasheet.
;
; Connections:
;
;  LCD Display       MSP430G2553
;    Pin #1            Ground
;    Pin #2            +5 volts
;    Pin #3            Contrast control (center pin from 10K potentiometer)
;    Pin #4            P1.4 (R/S Register Select)
;    Pin #5            Ground (Read/Write selection: 0 for Write, 1 for Read)
;    Pin #6            P1.5 (Enable)
;    Pin #7            P2.0 (Data)
;    Pin #8            P2.1 (Data)
;    Pin #9            P2.2 (Data)
;    Pin #10           P2.3 (Data)
;    Pin #11           P2.4 (Data)
;    Pin #12           P2.5 (Data)
;    Pin #13           P2.6 (Data)
;    Pin #14           P2.7 (Data)
;    Pin #15           +5 volts (backlight)
;    Pin #16           Ground (backlight)
;
; J. Kretzschmar


;-------------------------------------------------------------------------------
            .cdecls C,LIST,"msp430.h"      ; Include device header file
            .def    RESET
            .text
            .retain
            .retainrefs
;-------------------------------------------------------------------------------
RESET       mov.w   #__STACK_END,SP        ; Initialize stackpointer
StopWDT     mov.w   #WDTPW|WDTHOLD,&WDTCTL  ; Stop watchdog timer
;-------------------------------------------------------------------------------
;----- INITIAL SETUP MSP430G2553 PARAMETERS ------------------------------------
;-------------------------------------------------------------------------------
START_HERE:
    mov.b   &CALBC1_16MHZ,&BCSCTL1  ; Calibrated 16 MHz CPU Clock
    mov.b   &CALDCO_16MHZ,&DCOCTL   ; Calibrated 16 MHz CPU Clock

    mov.b   #00000000b, &P1SEL      ; Port 1 selected as I/O pins
    mov.b   #00000000b, &P2SEL      ; Port 2 selected as I/O pins
    mov.b   #00110000b, &P1DIR      ; P1.4 and P1.5 designated as OUTPUT pins
    mov.b   #11111111b, &P2DIR      ; P2.0 - P.7 designated as OUTPUT pins
    mov.b   #00000000b, &P1OUT      ; All Port_1 pins are cleared
    mov.b   #00000000b, &P2OUT      ; All Port_2 pins are cleared
;-------------------------------------------------------------------------------
;----- WAKE UP SEQUENCE OF LCD DISPLAY --- INITIAL 50 MILLISECOND DELAY ---------
;-------------------------------------------------------------------------------
    mov     #65535, R6             ; 12.5 Millisecond Delay
DELAY_0:
    dec     R6
    jnz     DELAY_0
    mov     #65535, R6             ; 12.5 Millisecond Delay
DELAY_00:
    dec     R6
```

```
      jnz     DELAY_00
      mov     #65535, R6          ; 12.5 Millisecond Delay
DELAY_000:
      dec     R6
      jnz     DELAY_000
      mov     #65535, R6          ; 12.5 Millisecond Delay
DELAY_0000:
      dec     R6
      jnz     DELAY_0000
;----- (COMMAND: FIRST WAKE UP CALL --- SELECTS 8-BIT, 2 LINE) ------------------
      mov.b   #00111000b, R4      ; 8-Bit entry format, 2 Line
      call    #SETUP
;----- (COMMAND: SECOND WAKE UP CALL --- SELECTS 8-BIT, 2 LINE) ----------------
      mov.b   #00111000b, R4      ; 8-Bit entry format, 2 Line
      call    #SETUP
;----- (COMMAND: THIRD WAKE UP CALL --- SELECTS 8-BIT, 2 LINE) ------------------
      mov.b   #00111000b, R4      ; 8-Bit entry format, 2 Line
      call    #SETUP
;-------------------------------------------------------------------------------
;----- SETUP INFORMATION TO BE ENTERED INTO THE LCD DISPLAY --------------------
;-------------------------------------------------------------------------------
;----- (COMMAND: CLEAR DISPLAY) ------------------------------------------------
      mov.b   #00000001b, R4      ; Clear Display
      call    #SETUP
;----- (COMMAND: SETS DDRAM ADDRESS --- WHERE CHARACTERS APPEAR) ----------------
      mov.b   #10000101b, R4      ; DDRAM Address 05 selected
      call    #SETUP
;----- (COMMAND: DISPLAY CONTROL) ----------------------------------------------
      mov.b   #00001100b, R4      ; Display ON, Cursor OFF, Cursor Blinking OFF
      call    #SETUP
;----- (COMMAND: DISPLAY SHIFT/CURSOR MOVE (RIGHT OR LEFT) ----------------------
      mov.b   #00000110b, R4      ; Display Shift OFF, Next character to the right
      call    #SETUP
;----- (COMMAND: CURSOR/DISPLAY SHIFT)(NOT IMPLEMENTED) ------------------------
      ;mov.b  #00010000b, R4      ; Cursor/Display Shift OFF, Shift Left
      ;call   #SETUP
;----- (COMMAND: CURSOR HOME)(NOT IMPLEMENTED) ---------------------------------
      ;mov.b  #00000010b, R4      ; Cursor Home (DDRAM address 00)
      ;call   #SETUP
      jmp     GOTO_DATA_ENTRY     ; All done with setup, goto Data Entry
;-------------------------------------------------------------------------------
;----- SUBROUTINE THAT ENTERS SETUP INFORMATION INTO LCD DISPLAY ----------------
;-------------------------------------------------------------------------------
SETUP:
      mov.b   #00000000b, &P1OUT  ; (P1.4 = 0) LCD DISPLAY Command Mode Selected

      bis.b   #00100000b, &P1OUT  ; (P1.5) ENABLE pin brought HIGH

      mov     #26215, R6          ; 5 Millisecond Delay
DELAY_1:
      dec     R6
      jnz     DELAY_1

      mov.b   R4, &P2OUT          ; Setup Information loaded into LCD DISPLAY

      bic.b   #00100000b, &P1OUT  ; (P1.5) ENABLE pin brought LOW

      mov     #26215, R6          ; 5 Millisecond Delay
DELAY_2:
      dec     R6
      jnz     DELAY_2

      ret                         ; Returns program to after "call"
```

```
;-----------------------------------------------------------------------------
;----- DATA (ASCII NUMBERS) TO BE ENTERED INTO LCD DISPLAY ----------------------
;-----------------------------------------------------------------------------
GOTO_DATA_ENTRY:
    mov     #77, R4             ; Loads ASCII number for "M"
    call    #ENTER_ASCII

    mov     #83, R4             ; Loads ASCII number for "S"
    call    #ENTER_ASCII

    mov     #80, R4             ; Loads ASCII number for "P"
    call    #ENTER_ASCII

    mov     #52, R4             ; Loads ASCII number for "4"
    call    #ENTER_ASCII

    mov     #51, R4             ; Loads ASCII number for "3"
    call    #ENTER_ASCII

    mov     #48, R4             ; Loads ASCII number for "0"
    call    #ENTER_ASCII
END_PROGRAM:
    jmp     END_PROGRAM         ; Program ends
;-----------------------------------------------------------------------------
;----- SUBROUTINE THAT ENTERS DATA (ASCII NUMBERS) INTO LCD DISPLAY -------------
;-----------------------------------------------------------------------------
ENTER_ASCII:
    mov.b   #00010000b, &P1OUT  ; (P1.4 = 1) LCD DISPLAY Data Mode Selected

    bis.b   #00100000b, &P1OUT  ; (P1.5) ENABLE pin brought HIGH

    mov     #26215, R6          ; 5 Millisecond Delay
DELAY_3:
    dec     R6
    jnz     DELAY_3

    mov.b   R4, &P2OUT          ; ASCII Numbers (Data) loaded into LCD Display

    bic.b   #00100000b, &P1OUT  ; P1.5 ENABLE pin brought LOW

    mov     #26215, R6          ; 5 Millisecond Delay
DELAY_4:
    dec     R6
    jnz     DELAY_4

    ret                         ; Returns program to after "call"
;-----------------------------------------------------------------------------
; Stack Pointer definition
;-----------------------------------------------------------------------------
            .global __STACK_END
            .sect   .stack
;-----------------------------------------------------------------------------
; Interrupt Vectors
;-----------------------------------------------------------------------------
            .sect   ".reset"            ; MSP430 RESET Vector
            .short  RESET
;-----------------------------------------------------------------------------
```

5.4   Task B

1. Connect the 16x2 LCD display to the MSP430G2553 microcontroller following the guide shown in Fig. 5.2. In this configuration you use the code listing "LCD_DISPLAY_PORT_2." Remember the two devices need to have a common ground. Wiring details are provided in Fig. 5.2.
2. Once you have the display working (displaying "MSP430") then implement the new lines of code you wrote in the PRELAB to display your first name.
3. Now change the code to shift the display three spaces to the left.
4. Take pictures of both of these changes to document your results.

Ground	Pin #1	WIRED CONNECTION	Pin #20	(Ground)
+5.0 Volts	Pin #2	See Notes Below		
Contrast Control	Pin #3	See Notes Below		
R/S Register Select	Pin #4	WIRED CONNECTION	Pin #6	(P1.4)
R/W Read - Write	Pin #5	See Notes Below		
Enable	Pin #6	WIRED CONNECTION	Pin #7	(P1.5)
Data-0	Pin #7	WIRED CONNECTION	Pin #8	(P2.0)
Data-1	Pin #8	WIRED CONNECTION	Pin #9	(P2.1)
Data-2	Pin #9	WIRED CONNECTION	Pin #10	(P2.2)
Data-3	Pin #10	WIRED CONNECTION	Pin #11	(P2.3)
Data-4	Pin #11	WIRED CONNECTION	Pin #12	(P2.4)
Data-5	Pin #12	WIRED CONNECTION	Pin #13	(P2.5)
Data-6	Pin #13	WIRED CONNECTION	Pin #19	(P2.6)
Data-7	Pin #14	WIRED CONNECTION	Pin #18	(P2.7)
Backlight (+5.0 v)	Pin #15			
Backlight (Ground)	Pin #16			
LCD DISPLAY				**MSP430G2553**

Notes:
- Pin #2 and Pin #15 - +5.0 volts (Use MSP430G2553 +5.0 volt supply)
- Pin #3 - Connect to wiper (middle terminal of a 10K pot)
- Pin #5 - Ground for writing to LCD Display

Fig. 5.2 16x2 LCD display with the MSP430G2553 connections

5.5   Post Lab Activities

1. In the Appendices we provide a brief introduction to Unified Modeling Language (UML) activity diagrams. These are UML compliant flow charts. Provide UML activity diagrams for programs provided in this laboratory exercise.
2. For assembly language programs provided; rewrite in C, compile, execute, and test for proper operation.

Lab 6: MSP430G2553 ADC Module

6

6.1 Purpose

This lab introduces the analog–to–digital converter module on the MSP430G2553 micro-controller. Analog–to–digital converters (ADC) provide a means to allow for interfacing real world signals with digital computers. Analog signals are continuously changing and are not simply "ON" or "OFF" like a digital signal. A few examples of analog signals are temperature, sound, or acceleration. Sensors sample analog signals and provide a small voltage that represents the signal at a particular moment in time. ADCs evaluate that voltage as a percentage of a reference voltage and output a binary number. The more bits the ADC uses to represent each sample, the better the original signal can be represented digitally. The ADC peripheral in the MSP430G2553 microcontroller consists of 8 ADC channels. Each channel can represent an inputted voltage with a 10–bit binary number. The inputted voltage is assigned a binary number somewhere between 0 and 1023.

Part one of this lab takes an analog voltage from a potentiometer and inputs this voltage into one of the channels of the ADC module. As the potentiometer passes through a specific voltage that is defined in the program code a LED will illuminate. In part two of the lab the same ADC channel is used, however, the 10–Bit number that represents the voltage is displayed on the LCD display presented in Lab #5.

6.2 Discussion

ADCs work by taking an input voltage, evaluating that voltage as a percentage of a reference voltage, and assigning it a "digital" number. The reference voltage used with the ADC in this lab is the supply voltage for the MSP430G2553, 3.3 V. For advance applications other reference voltages can be used.

© The Author(s), under exclusive license to Springer Nature Switzerland AG 2023   47
J. Kretzschmar et al., *MSP430 Microcontroller Lab Manual*, Synthesis Lectures on Digital
Circuits & Systems, https://doi.org/10.1007/978-3-031-26643-0_6

There are two parameters that are important when discussing ADCs: (1) What is the maximum number of bits that can be assigned by the ADC (common varieties are 8–bit, 10–bit, 12–bit, 14–bit, 16–bit ADCs) and, (2) What is the sample rate of the ADC? That is, how many times per second is the signal evaluated and assigned a digital number.

In the mid–1980s recorded music started to be stored (and sold) on compact discs (CDs). The music was sampled 44,000 times per second and then stored digitally on the compact disc. Replaying the music is exceptionally smooth and sounds great. Now consider if that same music was only sampled at a rate of 50 times per second. The music would sound very choppy and not pleasing to listen to. The reason for this poor reproduction is because of the slow sample rate overlooking substantial portions of the signal.

Similarly, if you were sampling the signal adequately, but, only assigning a four bit number to each sample then the binary number representing each sample would not provide very much resolution between sample points.

In summary, the faster the sample rate, and the higher number of bits of the digital number representing the signal, the better the sampled signal will be able to be digitally reproduced.

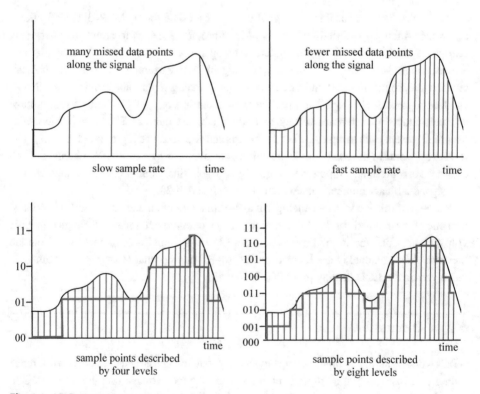

Fig. 6.1 ADC parameters

The importance of ADCs is that it allows analog signals to be represented by digital numbers, and then these numbers (data) can be manipulated and processed. This is the basis of digital signal processing (filters etc.). Figure 6.1 summarizes the important parameters of sample rate and ADC resolution (number of bits).

6.3   Part 1: ADC Demonstration with LED (C Code)

The MSP430G2553 LaunchPad and 5K Okm potentiometer shown in Fig. 6.2 are connected as follows: (1) Middle connection of potentiometer connected to P1.7, (2) One side of the potentiometer is connected to 3.3 V and, (3) The other side of the potentiometer is connected to GND. This is a voltage divider configuration that allows the input voltage to the ADC to be varied. Load the C code program "ADC_TEST" and slowly rotate the potentiometer. At some point in the rotation of the potentiometer the GREEN LED illuminates. Change the number "720" in the code listing to a number between 0–1023 and observe the result of the change.

Fig. 6.2 ADC demonstration

```
//-------------------------------------------------------------------
// Program Name: ADC_TEST
// C Code using Texas Instruments Code Composer
// Parts:
// -- Texas Instruments MSP-EXP430G2ET Launchpad with MSP430G2553
// -- 5K potentiometer
//
// Documentation:
// -- Texas Instruments SLAU144K Users Guide (ti.com)
// -- Texas Instruments SLAS735J MSP430G2553 Datasheet (ti.com)
//
// This program demonstrates how one channel of the ADC module functions.
// A potentiometer (variable resistor) is connected to the LaunchPad
// One end to 3.3 volts, the other end to Ground, and the center
// connection to pin P1.7.  As you rotate the potentiometer, at some point
// through its travels the GREEN LED on pin P1.0 will flash. Rotate slowly
// or else you may miss the flash. When the green LED flashes the ADC channel
// is outputting the number 720 which corresponds to the value defined in the
// code (could be any number 0-1023).
//
// Connections:
//
//      MSP430G2553
//          P1.7 (center pin from 5K potentiometer)
//          Ground (one end of potentiometer)
//          +3.3 volts (one end of potentiometer)
//
// J. Kretzschmar

#include <msp430.h>
int VALUE; // Integer variable "VALUE" declared
int DELAY; // Integer variable "DELAY" declared

void main(void)
{
    WDTCTL = WDTPW | WDTHOLD; // Stop watchdog timer
    BCSCTL1 = CALBC1_1MHZ;     // Calibrated 1 MHz clock
    DCOCTL = CALDCO_1MHZ;      // Calibrated 1 MHz clock
    P1DIR = BIT0;              // BIT0 is an output pin (GREEN LED)
    P1OUT = ~BIT0;             // BIT0 is a "0" (GREEN LED off)

// The next several lines set up the different aspects of the ADC.  The
// control registers are "ORed" with the aspect to be set up and then that
// is what is stored in the control register.  There are other ways to
// accomplish this task, however, this just keeps it understandable

    ADC10CTL1 = ADC10CTL1 | INCH_7;     // Pin P1.7 selected as input channel
    ADC10CTL0 = ADC10CTL0 | ADC10ON;    // ADC is turned on and the Sample
    ADC10CTL0 = ADC10CTL0 | ADC10SHT_2; // and Hold Time is extended long
                                        // enough to be able to see the flash

    while(1)        // This is a loop ... "while(TRUE)" keep looping
    {
        for(DELAY = 240; DELAY > 0; DELAY --); // The DELAY has the value of
                                               // 240 and as long as DELAY is
                                               // greater than 0 each time thru
                                               // it gets decremented by 1.
                                               // When 0 the program continues.
        ADC10CTL0 = ADC10CTL0 | ENC;           // ADC is enabled
        ADC10CTL0 = ADC10CTL0 | ADC10SC;       // ADC starts a converion
        for(DELAY = 240; DELAY > 0; DELAY --); // Same delay as above. Gives it
                                               // time to place the "VALUE" in
                                               // the holding register.
        VALUE = ADC10MEM;                      // A number between 0 and 1023 is
                                               // placed in the holding register
                                               // "ADC10MEM"
```

```
    if (VALUE == 720)                // If the "VALUE" equals 720 then
        (P1OUT = BIT0);              // turn "ON" the GREEN LED
    else                             // If the "VALUE" does not equal
        (P1OUT = P1OUT & ~BIT0);     // 720 then turn "OFF" the GREEN LED
    }                                // Keep looping and evaluating the
}                                    // "VALUE" in ADC10MEM

//-------------------------------------------------------------------
```

6.4 Part 2: Demonstration with LCD Display (Assembly Code)

Set up the MSP430G2553, LCD display as in LAB #5, and 5K potentiometer as in Part 1 of this lab as shown in Fig. 6.3. Load the assembly code program "ADC_LCD_PORT_2" and as the potentiometer is rotated numbers are displayed that represent the voltage inputted into the ADC. The numbers are a little jumpy, however, pick a number and using a voltmeter measure the voltage at the middle connection of the potentiometer.

Follow the example calculation provided below showing the correlation between the inputted voltage and the outputted digital number.

Fig. 6.3 ADC configuration

Example calculation:

$$720/1023 \quad Represents \quad (2.32\,\text{V})/(3.3\,\text{V})$$

$$2.32 = (3.3 \times 720)/1023$$

```
;------------------------------------------------------------------------
; Program Name: ADC_LCD_PORT_2
; Assembly Code using Texas Instruments Code Composer
; Parts:
; -- Texas Instruments MSP-EXP430G2ET Launchpad with MSP430G2553
; -- Newhaven NHD-0216K1Z 16x2 LCD Display (or similar display)
; -- 10K Potentiometer (for contrast control)
; -- 5K Potentiometer (for voltage divider for ADC input)
; Documentation:
; -- Texas Instruments SLAU144K Users Guide (ti.com)
; -- Texas Instruments SLAS735J MSP430G2553 Datasheet (ti.com)
; -- Newhaven NHD-0216K1Z Datasheet (newhavendisplay.com)
; -- Paper "Controlling a 16x2 LCD Display with the MSP430G2553 Microcontroller
;    Using Assembly Code" (Appendix)
;
; This program uses the ADC module on a MSP430G2553 microcontroller and displays
; the output number (somewhere between 0-1023) to a 16x2 LCD display for
; demonstration purposes. LCD display delays are longer than noted in the
; datasheet.
;
; Connections:
;
;  LCD Display      MSP430G2553
;    Pin #1           Ground
;    Pin #2           +5 volts
;    Pin #3           Contrast control (center pin from 10K potentiometer)
;    Pin #4           P1.4 (R/S Register Select)
;    Pin #5           Ground (Read/Write selection: 0 for Write, 1 for Read)
;    Pin #6           P1.5 (Enable)
;    Pin #7           P2.0 (Data)
;    Pin #8           P2.1 (Data)
;    Pin #9           P2.2 (Data)
;    Pin #10          P2.3 (Data)
;    Pin #11          P2.4 (Data)
;    Pin #12          P2.5 (Data)
;    Pin #13          P2.6 (Data)
;    Pin #14          P2.7 (Data)
;    Pin #15          +5 volts (backlight)
;    Pin #16          Ground (backlight)
;
;                     P1.7 (center pin from 5K potentiometer)
;                     Ground (one end of potentiometer)
;                     +3.3 volts (one end of potentiometer)
;
; J. Kretzschmar


;------------------------------------------------------------------------
            .cdecls C,LIST,"msp430.h"      ; Include device header file
            .def    RESET
            .text
            .retain
            .retainrefs
;------------------------------------------------------------------------
RESET       mov.w   #__STACK_END,SP         ; Initialize stackpointer
StopWDT     mov.w   #WDTPW|WDTHOLD,&WDTCTL   ; Stop watchdog timer
;------------------------------------------------------------------------
;----- INITIAL SETUP MSP430G2553 PARAMETERS -----------------------------
;------------------------------------------------------------------------
```

```
START_HERE:
    mov.b    &CALBC1_16MHZ,&BCSCTL1  ; Calibrated 16 MHz CPU Clock
    mov.b    &CALDCO_16MHZ,&DCOCTL   ; Calibrated 16 MHz CPU Clock

    mov.b    #00000000b, &P1SEL      ; Port 1 selected as I/O pins
    mov.b    #00000000b, &P2SEL      ; Port 2 selected as I/O pins
    mov.b    #00110000b, &P1DIR      ; P1.4 and P1.5 designated as OUTPUT pins
    mov.b    #11111111b, &P2DIR      ; P2.0 - P.7 designated as OUTPUT pins
    mov.b    #00000000b, &P1OUT      ; All Port_1 pins are cleared
    mov.b    #00000000b, &P2OUT      ; All Port_2 pins are cleared

    mov.w    #ADC10SHT_2+ADC10ON, &ADC10CTL0 ; Setting up ADC
    mov.w    #INCH_7, &ADC10CTL1             ; ADC input channel 7 (P1.7)

    bis.b    #BIT7, &ADC10AE0               ; Setting up ADC
;-----------------------------------------------------------------------------
;----- WAKE UP SEQUENCE OF LCD DISPLAY --- INITIAL 50 MILLISECOND DELAY --------
;-----------------------------------------------------------------------------
    mov      #65535, R6         ; 12.5 Millisecond Delay
DELAY_0:
    dec      R6
    jnz      DELAY_0
    mov      #65535, R6         ; 12.5 Millisecond Delay
DELAY_00:
    dec      R6
    jnz      DELAY_00
    mov      #65535, R6         ; 12.5 Millisecond Delay
DELAY_000:
    dec      R6
    jnz      DELAY_000
    mov      #65535, R6         ; 12.5 Millisecond Delay
DELAY_0000:
    dec      R6
    jnz      DELAY_0000
;----- (COMMAND: FIRST WAKE UP CALL --- SELECTS 8-BIT, 2 LINE) -----------------
    mov.b    #00111000b, R4     ; 8-Bit entry format, 2 Line
    call     #SETUP
;----- (COMMAND: SECOND WAKE UP CALL --- SELECTS 8-BIT, 2 LINE) ----------------
    mov.b    #00111000b, R4     ; 8-Bit entry format, 2 Line
    call     #SETUP
;----- (COMMAND: THIRD WAKE UP CALL --- SELECTS 8-BIT, 2 LINE) -----------------
    mov.b    #00111000b, R4     ; 8-Bit entry format, 2 Line
    call     #SETUP
;-----------------------------------------------------------------------------
;----- SETUP INFORMATION TO BE ENTERED INTO THE LCD DISPLAY ---------------------
;-----------------------------------------------------------------------------
;----- (COMMAND: CLEAR DISPLAY) -----------------------------------------------
    mov.b    #00000001b, R4     ; Clear Display
    call     #SETUP
;----- (COMMAND: SETS DDRAM ADDRESS --- WHERE CHARACTERS APPEAR) ---------------
    mov.b    #10000111b, R4     ; DDRAM Address 07 selected
    call     #SETUP
;----- (COMMAND: DISPLAY CONTROL) ---------------------------------------------
    mov.b    #00001100b, R4     ; Display ON, Cursor OFF, Cursor Blinking OFF
    call     #SETUP
;----- (COMMAND: DISPLAY SHIFT/CURSOR MOVE (RIGHT OR LEFT) ---------------------
    mov.b    #00000110b, R4     ; Display Shift OFF, Next character to the right
    call     #SETUP
;----- (COMMAND: CURSOR/DISPLAY SHIFT)(NOT IMPLEMENTED) ------------------------
    ;mov.b   #00010000b, R4     ; Cursor/Display Shift OFF, Shift Left
    ;call    #SETUP
;----- (COMMAND: CURSOR HOME)(NOT IMPLEMENTED) --------------------------------
    ;mov.b   #00000010b, R4     ; Cursor Home (DDRAM address 00)
    ;call    #SETUP
    jmp      GOTO_DATA_ENTRY    ; All done with setup, goto Data Entry
;-----------------------------------------------------------------------------
;----- SUBROUTINE THAT ENTERS SETUP INFORMATION INTO LCD DISPLAY ----------------
```

```
;-------------------------------------------------------------------------------
SETUP:
    mov.b   #00000000b, &P1OUT  ; (P1.4 = 0) LCD DISPLAY Command Mode Selected

    bis.b   #00100000b, &P1OUT  ; (P1.5) ENABLE pin brought HIGH

    mov     #26215, R6          ; 5 Millisecond Delay
DELAY_1:
    dec     R6
    jnz     DELAY_1

    mov.b   R4, &P2OUT          ; Setup Information loaded into LCD DISPLAY

    bic.b   #00100000b, &P1OUT  ; (P1.5) ENABLE pin brought LOW

    mov     #26215, R6          ; 5 Millisecond Delay
DELAY_2:
    dec     R6
    jnz     DELAY_2

    ret                         ; Returns program to after "call"
;-------------------------------------------------------------------------------
;----- DATA FROM ADC ACQUIRED -- CONVERTED TO BCD NUMBER -- THEN CONVERTED TO
;----- ASCII NUMBERS AND THEN SENT TO THE LCD DISPLAY -------------------------
;-------------------------------------------------------------------------------
GOTO_DATA_ENTRY:

LOOP:
    mov     #30, R14    ; Delay routine. At a 16 MHz clock speed the
DELAY_3:                ; numbers come in and get processed so fast that
    mov     #65535, R6  ; the LCD jumps around too much. The value of
DELAY_4:                ; "30" can be changed to slow down or speed up
    dec     R6          ; depending on the application
    jnz     DELAY_4
    dec     R14
    jnz     DELAY_3

    mov.b   #10000111b, R4 ; This clears out the old number and
    call    #SETUP         ; re-establishes where the new number will be
    mov     #525, R6       ; displayed (DDRAM Address 07)
DELAY_5:
    dec     R6
    jnz     DELAY_5
;-------------------------------------------------------------------------------
;----- ACQUIRING DATA FROM ADC -------------------------------------------------
;-------------------------------------------------------------------------------
    bis.w   #ENC+ADC10SC, &ADC10CTL0 ; Enables and Starts a conversion
    mov.w   #20000, R8      ; Delay routine to give time to settle
DELAY_6:                    ; before transferring data
    dec     R8
    jnz     DELAY_6
    mov.w   &ADC10MEM, R9   ; Transfer data from ADC to R9
;-------------------------------------------------------------------------------
;----- CONVERTING ADC NUMBER TO BCD NUMBER -------------------------------------
;-------------------------------------------------------------------------------
    mov     #0, R15     ; Make sure R15 is empty
    rla.w   R9          ; This section converts the binary number
    dadd.w  R15, R15    ; from the ADC into a BCD number. The
    rla.w   R9          ; assembly statement "dadd" that does this is
    dadd.w  R15, R15    ; only available for use when programming in
    rla.w   R9          ; assembly. These lines take the 16 Bit register
    dadd.w  R15, R15    ; and roll off one Bit at a time to the left
    rla.w   R9          ; and the "BCD Conversion" process is accomplished
    dadd.w  R15, R15    ; with "dadd" and stored in R6. These lines could
    rla.w   R9          ; be done with a loop process (x16), however it is
    dadd.w  R15, R15    ; important to see how the BCD conversion is done
    rla.w   R9
```

```
            dadd.w      R15, R15
            rla.w       R9
            dadd.w      R15, R15
            rla.w       R9
            dadd.w      R15, R15
            rla.w       R9
            dadd.w      R15, R15
            rla.w       R9
            dadd.w      R15, R15
            rla.w       R9
            dadd.w      R15, R15
            rla.w       R9
            dadd.w      R15, R15
            rla.w       R9
            dadd.w      R15, R15
            rla.w       R9
            dadd.w      R15, R15
            rla.w       R9
            dadd.w      R15, R15
            rla.w       R9
            dadd.w      R15, R15
;-------------------------------------------------------------------------
;----- READS BCD NUMBER IN BLOCKS OF 4 BITS ------------------------------
;-------------------------------------------------------------------------
            mov         #0, R7      ; Ensures R7 is empty
            mov         #4, R10     ; Place 4 in counter R10
READ_4_BITS:                        ; This routine takes the BCD number stored in R15
            rlc         R15         ; and evaluates each block of 4 Bits one bit at a
            jc          EIGHT       ; time starting from the left. R7 stores the number
ROLL_BIT2:                          ; which is then converted to the appropriate ASCII
            rlc         R15         ; number and sent to be displayed on the LCD display
            jc          FOUR        ; This process is done 4 times so that the entire
ROLL_BIT1:                          ; 16 Bit register (R15) is read (one BCD number per
            rlc         R15         ; 4 Bit block) ... then the process repeats with a
            jc          TWO         ; new number from the ADC
ROLL_BIT0:
            rlc         R15
            jc          ONE
            jmp         DIGIT


EIGHT:
            add         #8, R7
            jmp         ROLL_BIT2
FOUR:
            add         #4, R7
            jmp         ROLL_BIT1
TWO:
            add         #2, R7
            jmp         ROLL_BIT0
ONE:
            add         #1, R7
;-------------------------------------------------------------------------
;----- TAKES BCD DIGIT AND CONVERTS IT TO ASCII NUMBER -------------------
;-------------------------------------------------------------------------
DIGIT:
            add         #48, R7     ; ASCII 48 = 0   Refer to ASCII table
            mov         R7, R4

            call        #ENTER_ASCII ; Sends ASCII number to be displayed

            mov         #0, R4      ; R4 cleared
            mov         #0, R7      ; R7 cleared
            dec         R10         ; 1 decremented from R10
            cmp         #0, R10     ; Does R10 have 0 stored in it
            jeq         LOOP        ; If yes, go get new ADC number
            jmp         READ_4_BITS ; If no, go get new 4-bit block
;-------------------------------------------------------------------------
```

```
;----- SUBROUTINE THAT ENTERS ADC DATA (ASCII NUMBERS) INTO LCD DISPLAY ---------
;-------------------------------------------------------------------------------
ENTER_ASCII:
    mov.b   #00010000b, &P1OUT  ; (P1.4 = 1) LCD DISPLAY Data Mode Selected

    bis.b   #00100000b, &P1OUT  ; (P1.5) ENABLE pin brought HIGH

    mov     #26215, R6          ; 5 Millisecond Delay
DELAY_7:
    dec     R6
    jnz     DELAY_7

    mov.b   R4, &P2OUT          ; ASCII Numbers (Data) loaded into LCD Display

    bic.b   #00100000b, &P1OUT  ; P1.5 ENABLE pin brought LOW

    mov     #26215, R6          ; 5 Millisecond Delay
DELAY_8:
    dec     R6
    jnz     DELAY_8

    ret                         ; Returns program to after "call"
;-------------------------------------------------------------------------------
; Stack Pointer definition
;-------------------------------------------------------------------------------
            .global __STACK_END
            .sect   .stack
;-------------------------------------------------------------------------------
; Interrupt Vectors
;-------------------------------------------------------------------------------
            .sect   ".reset"            ; MSP430 RESET Vector
            .short  RESET
;-------------------------------------------------------------------------------
```

6.5   Post Lab Activities

1. In the Appendices we provide a brief introduction to Unified Modeling Language (UML) activity diagrams. These are UML compliant flow charts. Provide UML activity diagrams for programs provided in this laboratory exercise.

2. For assembly language programs provided; rewrite in C, compile, execute, and test for proper operation.

Lab 7: Timer_A Module of the MSP430G2553 (Introduction to Interrupts)

7.1 Purpose

This lab introduces the Timer_A module of the MSP430G2553 which is a peripheral used in developing pulse width modulation signals, precision timing, and for measurement applications.

The Timer_A module has two separate 16–bit timers, Timer_0 and Timer_1. Each Timer can count from 0–65535, and each timer has three capture/compare registers available for use. Each timer has timer count modes that can be configured for capture and compare applications: the Up Mode, the Continuous Mode, and the Up/Down Mode. Each of these modes are discussed in detail. In addition to learning about the Timer_A module, interrupts are also introduced and incorporated into the Timer_A module code examples.

7.2 Discussion

Understanding how to use the Timers on the MSP430G2553 can be confusing as the naming of the registers that control the functions may be difficult to interpret from the documentation. For example, TACTL is the Timer_A Control Register and allows selection of the TAR clock to drive the timer, and other parameters. After reviewing the Header file which defines parameters for the MSP430 family of microcontrollers, TACTL is associated with the Timer_0 timer. It is implied that TACTL is actually TA0CTL, and program statements compile with either TACTL or TA0CTL. If using Timer_1 the instruction must be written as TA1CTL. To avoid frustration when writing code, add in the "0s" and "1s" in the register names for the timer used.

© The Author(s), under exclusive license to Springer Nature Switzerland AG 2023
J. Kretzschmar et al., *MSP430 Microcontroller Lab Manual*, Synthesis Lectures on Digital
Circuits & Systems, https://doi.org/10.1007/978-3-031-26643-0_7

Both of the timers in the Timer_A module use the TAR clock as the counting base. Since each timer is 16–bits it can count from 0–65535, and the rate at which the timer counts from 0–65535 is dependent on the clock speed. If the TAR clock is set for 1 MHz then the time it would take to count from 0–65535 would be 65.535 ms. Similarly, if the TAR clock was set for 16 MHZ then the time would be 4.095 ms. In order to better understand, and visually see what is happening on an oscilloscope, when the capture and compare functions are engaged, the clock speed in the code listings is set for approximately 12 KHz (the VLOCLK).

When using Timer_0 and Timer_1, each timer has three Capture/Compare registers used to set parameters, and to store sampled data. In the code listings for this lab only TA0CCR0 is used.

7.3   Capture/Compare Registers of Timer_0 and Timer_1

- TA0CCR0: Timer_0 Capture/Compare Register 0
- TA0CCR1: Timer_0 Capture/Compare Register 1
- TA0CCR2: Timer_0 Capture/Compare Register 2
- TA1CCR0: Timer_1 Capture/Compare Register 0
- TA1CCR1: Timer_1 Capture/Compare Register 1
- TA1CCR2: Timer_1 Capture/Compare Register 2.

7.4   Up Mode

The Up Mode allows the programmer to choose the upper counting limit at which the counter rolls around to zero and starts counting again, see Fig. 7.1. This is useful for setting a defined period of time, and the set point is designated by TA0CCR0 in the TIMER_UPMODE code listing. The program TIMER_UPMODE generates a pulse when the timer reaches TA0CCR0, this pulse is outputted to pin P1.6 and is visualized on the oscilloscope, see Figs. 7.2, 7.3, and 7.4.

7.4.1   TIMER_UPMODE Code Listing

This code listing sets up the Timer_0 to operate in the Up Mode and allows for changing the value of TA0CCR0 which produces different time intervals between pulses that are outputted on pin P1.6. To see these changes the output on pin P1.6 is connected to an oscilloscope. Initially, TA0CCR0 is set at 5000, and the lines of code with TA0CCR0 set at 10000 and 20000 are commented out. Note that with the TAR clock set at a slow rate, entering in a large number for TA0CCR0, results in a long time interval between pulses.

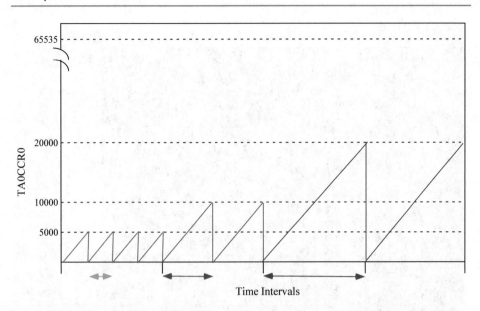

Fig. 7.1 Up mode

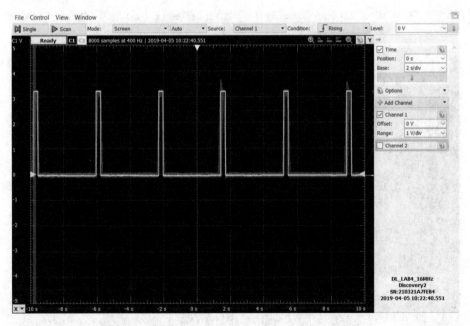

Fig. 7.2 TA0CCR0:5000 (approximately 3.5 s interval between pulses) TAR Clock = VLOCLK (approximately 12 KHz)

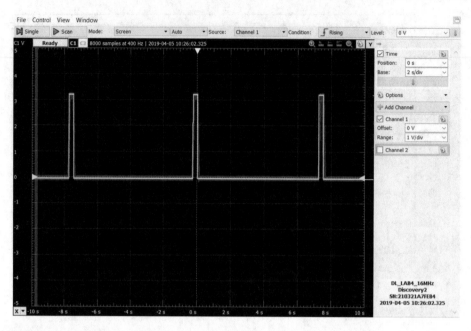

Fig. 7.3 TA0CCR0:10000 (approximately 7 s interval between pulses) TAR Clock = VLOCLK (approximately 12 KHz)

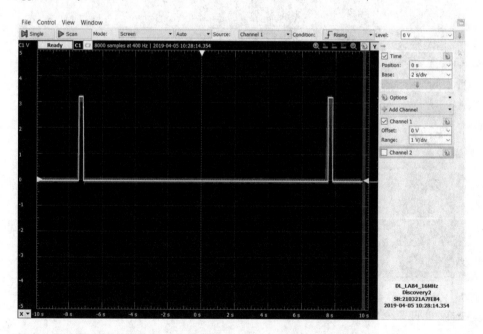

Fig. 7.4 TA0CCR0: 20000 (approximately 15 s interval between pulses) TAR Clock = VLOCLK (approximately 12 KHz)

```
;------------------------------------------------------------------------

; Program Name: TIMER_UPMODE
; Assembly Code using Texas Instruments Code Composer
; Parts:
; -- Texas Instruments MSP-EXP430G2ET Launchpad with MSP430G2553
; -- Oscilloscope
; Documentation:
; -- Texas Instruments SLAU144K Users Guide (ti.com)
; -- Texas Instruments SLAS735J MSP430G2553 Datasheet (ti.com)
;
; This program demonstrates the use of the Timer in the Up Mode where TA0CCR0 is
; set to the level you select. The timer counts up to that level which triggers
; an interrupt, generates a pulse on pin P1.6 (turns RED LED on) then drops back
; to zero and starts again. The CPU clock drives the rate at which the timer
; counts. A very slow clock is used so that you can see the time intervals
; when different values are entered in for TA0CCR0. The RED LED (P1.6) is turned
; on for a short time as the Interupt Service Routine is run. Connect P1.6 to an
; oscilloscope. Approximate example time intervals between pulses on P1.6 are:
; (1) TA0CCR0 = 20000 [approx 15 second interval], (2) TA0CCR0 = 10000 [approx
; 7 second interval], (3) TA0CCR0 = 5000 [approx 3 second interval]. Uncomment
; the different TA0CCR0 values (only one at a time) and look at the generated
; pulses outputted on P1.6 on an oscilloscope.
;
; Connections:
;
;    MSP430G2553            Oscilloscope
;       P1.6  ---------->  Input
;       Ground             Ground
;
; J. Kretzschmar
;
;------------------------------------------------------------------------
            .cdecls C,LIST,"msp430.h"      ; Include device header
            .def    RESET
            .text
            .retain
            .retainrefs
;------------------------------------------------------------------------
RESET       mov.w   #__STACK_END,SP        ; Initialize stackpointer
StopWDT     mov.w   #WDTPW|WDTHOLD,&WDTCTL  ; Stop watchdog timer
;------------------------------------------------------------------------
;    SLOW CLOCK SETUP (VLOCLK - APPROX 12 KHZ) FOR THE TIMER
;------------------------------------------------------------------------
    mov.b       #00000000b, &BCSCTL1  ; Basic Clock Control Register 1
    mov.b       #00100000b, &BCSCTL3  ; Bits 5+4 select the VLOCLK
    mov.b       #11111110b, &BCSCTL2  ; Bits 7+6 select the VLOCLK
                                      ; BITS 5+4 divide MCLK/8
                                      ; Bit 3 selects VLOCLK for SMCLK
                                      ; Bits 2+1 divide SMCLK/8
;------------------------------------------------------------------------
;----- SETTING UP TIMER -------------------------------------------------
;------------------------------------------------------------------------
    mov.w       #CCIE, &TA0CCTL0      ; TA0CCTL0 = Capture/Compare Register
                                      ; CCIE = Interupt Enabled
                                      ; Compare function is used
    mov.w       #TASSEL_2 | ID_0 | MC_1, &TA0CTL
                                      ; TA0CTL = Timer_0 Control Register
                                      ; TASSEL_2 = SMCLK selected for clock
                                      ; ID_0 = divide SMCLK/1
                                      ; MC_1 = Up Mode (counts to TACCR0)

;----- SELECT VALUE FOR TA0CCR0 ----- ONLY SELECT ONE AT A TIME ----------------
    mov.w       #5000, &TA0CCR0       ; Counts from 0-5000
    ;mov.w      #10000, &TA0CCR0      ; Counts from 0-10000
    ;mov.w      #20000, &TA0CCR0      ; Counts from 0-20000
;------------------------------------------------------------------------
```

```
        mov.b      #01000000b, &P1DIR     ; P1.6 assigned as output pin

LOOP:
        bis.w      #GIE, SR               ; Global Interupts Enabled
        jmp        LOOP                   ; Loops until interrupt is triggered
;-----------------------------------------------------------------------------
;----- INTERRUPT SERVICE ROUTINE ----- GENERATES A PULSE ON P1.6 ---------------
;-----------------------------------------------------------------------------
PULSE:                                    ; Interupt Service Routine

        mov        #500, R10              ; Delay: keeps RED LED on for short period
RED:
        bis.b      #01000000b, &P1OUT     ; Turns RED LED on (P1.6)
        dec        R10
        jnz        RED
        bic.b      #01000000b, &P1OUT     ; Turns RED LED off (P1.6)

        reti                              ; Return from interrupt
;-----------------------------------------------------------------------------
; Stack Pointer definition
;-----------------------------------------------------------------------------
        .global __STACK_END
        .sect   .stack
;-----------------------------------------------------------------------------
; Interrupt Vectors
;-----------------------------------------------------------------------------
;       .sect   ".int00"    ; Slot 0
;       .sect   ".int01"    ; Slot 1
;       .sect   ".int02"    ; Slot 2    Port1_Vector
;       .sect   ".int03"    ; Slot 3    Port2_Vector
;       .sect   ".int04"    ; Slot 4
;       .sect   ".int05"    ; Slot 5    ADC10_Vector
;       .sect   ".int06"    ; Slot 6    USCIAB0TX_Vector
;       .sect   ".int07"    ; Slot 7    USCIAB0RX_Vector
;       .sect   ".int08"    ; Slot 8    TIMER0_A1_Vector
        .sect   ".int09"    ; Slot 9    TIMER_0 Vector
        .short  PULSE
;       .sect   ".int10"    ; Slot 10   WDT_Vector
;       .sect   ".int11"    ; Slot 11   COMPARATORA_Vector
;       .sect   ".int12"    ; Slot 12   TIMER1_A1_Vector
;       .sect   ".int13"    ; Slot 13   TIMER1_A0_Vector
;       .sect   ".int14"    ; Slot 14   NMI_Vector
        .sect   ".reset"    ; Slot 15   MSP430 RESET Vector
        .short  RESET

;-----------------------------------------------------------------------------
```

7.5   Continuous Mode

The Continuous Mode Fig. 7.5 has a fixed point of 65535 that the timer counts up to before the counter rolls around to zero and starts counting again. This is useful for determining time intervals between events. The program TIMER_CAPTURE uses the timer in the continuous mode and when a tactile switch on pin P1.1 has been pushed an event is triggered. When the switch is pushed, the number at that moment is captured as the counter progresses from 0–65535. The captured number is stored in TA0CCR0 and then sent to an LCD display.

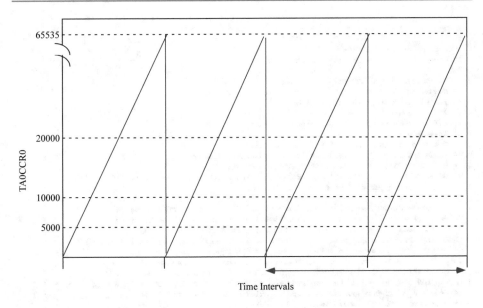

Fig. 7.5 Continuous mode

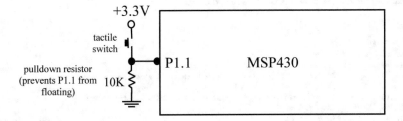

Fig. 7.6 Tactile switch connection

7.5.1 TIMER_CAPTURE Code Listing

The MSP430G2553 LaunchPad and the LCD display are configured as was done in Chaps. 5 and 6. A tactile switch is connected as shown in Fig. 7.6.

The 10K resistor is a pull down resistor so that input pin P1.1 is not floating. An internal Pull–down resistor could have been implemented instead of the external Pull–down resistor. In many applications inputs that would trigger an event would come from a sensor. Initially, the LCD display is blank, and when the tactile switch is pushed a number between 0–65535 appears (Fig. 7.7). Push the switch again and a new number appears. At 65535 the counter rolls over and starts counting again from 0.

```
;------------------------------------------------------------------------

; Program Name: TIMER_CAPTURE
; Assembly Code using Texas Instruments Code Composer
; Parts:
; -- Texas Instruments MSP-EXP430G2ET Launchpad with MSP430G2553
; -- Newhaven NHD-0216K1Z 16x2 LCD Display (or similar display)
; -- 10K Potentiometer (for contrast control)
; -- Tactile Switch
; -- 10K resistor for external pulldown resistor on pin P1.1
; Documentation:
; -- Texas Instruments SLAU144K Users Guide (ti.com)
; -- Texas Instruments SLAS735J MSP430G2553 Datasheet (ti.com)
; -- Newhaven NHD-0216K1Z Datasheet (newhavendisplay.com)
; -- Paper "Controlling a 16x2 LCD Display with the MSP430G2553 Microcontroller
;    Using Assembly Code" (Appendix)
;
; This program demonstrates the use of the Timer in the Capture Mode using an
; interupt generated by an outside pulse inputted into pin P1.1. The Timer is
; setup to operate in Continuous Mode which means it counts from 0-65535 and
; then drops back to 0 and starts counting again. The timer is driven by the
; CPU clock (the speed can be varied). Once the interupt has been received the
; program then goes to the Timer and "Captures" the count at that moment. This
; count is then stored in &TA0CCR0, transferred to R9 and processed for display
; on the LCD display. After the captured number has been displayed the program
; returns and waits for another interupt. A quick push of the tactile switch
; triggers an interrupt which displays a number (0-65535) on the LCD Display.
; The CPU has been purposefully slowed down so that multiple interupts can be
; accomplished and visualized as the Timer progresses from 0-65535.
; LCD display delays are longer than noted in the datasheet.
;
; Connections:
;
;   LCD Display      MSP430G2553
;     Pin #1           Ground
;     Pin #2           +5 volts
;     Pin #3           Contrast control (center pin from 10K potentiometer)
;     Pin #4           P1.4 (R/S Register Select)
;     Pin #5           Ground (Read/Write selection: 0 for Write, 1 for Read)
;     Pin #6           P1.5 (Enable)
;     Pin #7           P2.0 (Data)
;     Pin #8           P2.1 (Data)
;     Pin #9           P2.2 (Data)
;     Pin #10          P2.3 (Data)
;     Pin #11          P2.4 (Data)
;     Pin #12          P2.5 (Data)
;     Pin #13          P2.6 (Data)
;     Pin #14          P2.7 (Data)
;     Pin #15          +5 volts (backlight)
;     Pin #16          Ground (backlight)
;
;                      P1.1 (one pole of tactile switch)
;                      10K resistor from P1.1 to Ground (pulldown resistor)
;                      +3.3 volts (one pole of tactile switch)
;
; J. Kretzschmar

;------------------------------------------------------------------------
            .cdecls C,LIST,"msp430.h"      ; Include device header file
            .def    RESET
            .text
            .retain
            .retainrefs
;------------------------------------------------------------------------
RESET       mov.w   #__STACK_END,SP        ; Initialize stackpointer
StopWDT     mov.w   #WDTPW|WDTHOLD,&WDTCTL  ; Stop watchdog timer
;------------------------------------------------------------------------
```

```
;----- INITIAL SETUP MSP430G2553 PARAMETERS -------------------------------------
;--------------------------------------------------------------------------------
START_HERE:
    mov.b    &CALBC1_16MHZ,&BCSCTL1   ; Calibrated 16 MHz CPU Clock
    mov.b    &CALDCO_16MHZ,&DCOCTL    ; Calibrated 16 MHz CPU Clock

    mov.b    #00000010b, &P1SEL       ; P1.1 selected as TIMER_0 (TA.0) input
    mov.b    #00000000b, &P2SEL       ; Port 2 selected as I/O pins
    mov.b    #00110000b, &P1DIR       ; P1.4 and P1.5 designated as OUTPUT pins
    mov.b    #11111111b, &P2DIR       ; P2.0 - P.7 designated as OUTPUT pins
    mov.b    #00000000b, &P1OUT       ; All Port_1 pins are cleared
    mov.b    #00000000b, &P2OUT       ; All Port_2 pins are cleared
;--------------------------------------------------------------------------------
;----- WAKE UP SEQUENCE OF LCD DISPLAY --- INITIAL 50 MILLISECOND DELAY ---------
;--------------------------------------------------------------------------------
    mov      #65535, R6        ; 12.5 Millisecond Delay
DELAY_0:
    dec      R6
    jnz      DELAY_0
    mov      #65535, R6        ; 12.5 Millisecond Delay
DELAY_00:
    dec      R6
    jnz      DELAY_00
    mov      #65535, R6        ; 12.5 Millisecond Delay
DELAY_000:
    dec      R6
    jnz      DELAY_000
    mov      #65535, R6        ; 12.5 Millisecond Delay
DELAY_0000:
    dec      R6
    jnz      DELAY_0000
;----- (COMMAND: FIRST WAKE UP CALL --- SELECTS 8-BIT, 2 LINE) ------------------
    mov.b    #00111000b, R4    ; 8-Bit entry format, 2 Line
    call     #SETUP
;----- (COMMAND: SECOND WAKE UP CALL --- SELECTS 8-BIT, 2 LINE) -----------------
    mov.b    #00111000b, R4    ; 8-Bit entry format, 2 Line
    call     #SETUP
;----- (COMMAND: THIRD WAKE UP CALL --- SELECTS 8-BIT, 2 LINE) ------------------
    mov.b    #00111000b, R4    ; 8-Bit entry format, 2 Line
    call     #SETUP
;--------------------------------------------------------------------------------
;----- SETUP INFORMATION TO BE ENTERED INTO THE LCD DISPLAY ---------------------
;--------------------------------------------------------------------------------
;----- (COMMAND: CLEAR DISPLAY) -------------------------------------------------
    mov.b    #00000001b, R4    ; Clear Display
    call     #SETUP
;----- (COMMAND: SETS DDRAM ADDRESS --- WHERE CHARACTERS APPEAR) ----------------
    mov.b    #10000111b, R4    ; DDRAM Address 07 selected
    call     #SETUP
;----- (COMMAND: DISPLAY CONTROL) -----------------------------------------------
    mov.b    #00001100b, R4    ; Display ON, Cursor OFF, Cursor Blinking OFF
    call     #SETUP
;----- (COMMAND: DISPLAY SHIFT/CURSOR MOVE (RIGHT OR LEFT) ----------------------
    mov.b    #00000110b, R4    ; Display Shift OFF, Next character to the right
    call     #SETUP
;----- (COMMAND: CURSOR/DISPLAY SHIFT)(NOT IMPLEMENTED) -------------------------
    ;mov.b   #00010000b, R4    ; Cursor/Display Shift OFF, Shift Left
    ;call    #SETUP
;----- (COMMAND: CURSOR HOME)(NOT IMPLEMENTED) ---------------------------------
    ;mov.b   #00000010b, R4    ; Cursor Home (DDRAM address 00)
    ;call    #SETUP
    jmp      GOTO_DATA_ENTRY   ; All done with setup, goto Data Entry
;--------------------------------------------------------------------------------
;----- SUBROUTINE THAT ENTERS SETUP INFORMATION INTO LCD DISPLAY ----------------
;--------------------------------------------------------------------------------
SETUP:
    mov.b    #00000000b, &P1OUT  ; (P1.4 = 0) LCD DISPLAY Command Mode Selected
```

```
       bis.b    #00100000b, &P1OUT   ; (P1.5) ENABLE pin brought HIGH

       mov      #26215, R6           ; 5 Millisecond Delay
DELAY_1:
       dec      R6
       jnz      DELAY_1

       mov.b    R4, &P2OUT           ; Setup Information loaded into LCD DISPLAY

       bic.b    #00100000b, &P1OUT   ; (P1.5) ENABLE pin brought LOW

       mov      #26215, R6           ; 5 Millisecond Delay
DELAY_2:
       dec      R6
       jnz      DELAY_2

       ret                           ; Returns program to after "call"
;----------------------------------------------------------------------------
;----- DATA FROM TIMER ACQUIRED -- CONVERTED TO BCD NUMBER -- THEN CONVERTED TO
;----- ASCII NUMBERS AND THEN SENT TO THE LCD DISPLAY ------------------------
;----------------------------------------------------------------------------
GOTO_DATA_ENTRY:
;----------------------------------------------------------------------------
;---- SLOW CLOCK SETUP (VLOCLK - APPROX 12 KHZ) FOR THE TIMER ----------------
;----------------------------------------------------------------------------
       mov.b    #00000000b, &BCSCTL1    ; Basic Clock Control Register 1
       mov.b    #00100000b, &BCSCTL3    ; Bits 5+4 select the VLOCLK
       mov.b    #11111110b, &BCSCTL2    ; Bits 7+6 select the VLOCLK
                                        ; Bits 5+4 divide MCLK/8
                                        ; Bit 3 selects VLOCLK for SMCLK
                                        ; Bits 2+1 divide SMCLK/8

       mov.w    #CM_1 | CCIS_0 | CAP | CCIE | SCS, &TA0CCTL0
                                  ; TA0CCTL0 = Capture + Control Register "0"
                                  ; CM_1 = Capture on rising edge
                                  ; CCIS_0 = CCI0A Capture Input Select (P1.1)
                                  ; CAP = Capture Mode
                                  ; CCIE = Interupt Enabled
                                  ; SCS = Synchronous Capture (always use this)
       mov.w    #TASSEL_2 | MC_2, &TA0CTL
                                  ; TA0CTL = Timer "0" Control Register
                                  ; TASSEL_2 = SMCLK selected for the clock
                                  ; MC_2 = Continuous Mode (counts 0-65535)

       bis.w    #GIE | LPM1, SR       ; Global Interupts Enabled, Low Power Mode
;----------------------------------------------------------------------------
;---- INTERUPT SERVICE ROUTINE ----------------------------------------------
;----------------------------------------------------------------------------
CAPTURE_ONE:
       mov.w    &TA0CCR0, R9          ; Captured number stored in TA0CCR0 sent to R9

       mov.b    &CALBC1_16MHZ, &BCSCTL1 ; Fast CPU clock for setting up the LCD
       mov.b    &CALDCO_16MHZ, &DCOCTL  ; Fast CPU clock for setting up the LCD

       mov.b    #10000111b, R4        ; Old number cleared, re-establish location
       call     #SETUP                ; Goto SETUP

       mov      #26215, R6            ; 5 millisecond delay
DELAY_3:                              ; Keeps LCD Display from being too jumpy
       dec      R6
       jnz      DELAY_3
;----------------------------------------------------------------------------
;---- ROUTINE THAT TAKES BINARY NUMBER STORED IN R9 AND CONVERTS TO BCD ------
;---- NEW "BCD" NUMBER STORED IN R15 AND R5. IT TAKES 5 BLOCKS OF 4 BITS -----
;---- TO STORE A 5 DIGIT BCD NUMBER -----------------------------------------
;----------------------------------------------------------------------------
```

```
        clr     R15
        clr     R5
        rla.w   R9
        dadd.w  R15, R15
        dadd.w  R5, R5
        rla.w   R9
        dadd.w  R15, R15
        dadd.w  R5, R5
        rla.w   R9
        dadd.w  R15, R15
        dadd.w  R5, R5
        rla.w   R9
        dadd.w  R15, R15
        dadd.w  R5, R5
        rla.w   R9
        dadd.w  R15, R15
        dadd.w  R5, R5
        rla.w   R9
        dadd.w  R15, R15
        dadd.w  R5, R5
        rla.w   R9
        dadd.w  R15, R15
        dadd.w  R5, R5
        rla.w   R9
        dadd.w  R15, R15
        dadd.w  R5, R5
        rla.w   R9
        dadd.w  R15, R15
        dadd.w  R5, R5
        rla.w   R9
        dadd.w  R15, R15
        dadd.w  R5, R5
        rla.w   R9
        dadd.w  R15, R15
        dadd.w  R5, R5
        rla.w   R9
        dadd.w  R15, R15
        dadd.w  R5, R5
        rla.w   R9
        dadd.w  R15, R15
        dadd.w  R5, R5
        rla.w   R9
        dadd.w  R15, R15
        dadd.w  R5, R5
        rla.w   R9
        dadd.w  R15, R15
        dadd.w  R5, R5
        rla.w   R9
        dadd.w  R15, R15
        dadd.w  R5, R5
;-------------------------------------------------------------------------------
; THIS ROUTINE REMOVES THE UPPER 12 BITS OF R5 AND READS THE LOWER 4 BITS OF R5
; (First decimal digit of the BCD number) AND CREATES AN ASCII NUMBER THAT IS
; SENT TO THE LCD DISPLAY ------------------------------------------------------
;-------------------------------------------------------------------------------
        mov     #12, R10        ; Counter
        mov     #0, R11         ; Ensure R11 has 0 in it (Accumulator)
ROLL_12:
        cmp     #0, R10
        jeq     ROLL_BIT_3
        dec     R10
        rla     R5
        jnz     ROLL_12
ROLL_BIT_3:
        rla     R5
        jc      EIGHT
ROLL_BIT_2:
```

```
        rla     R5
        jc      FOUR
ROLL_BIT_1:
        rla     R5
        jc      TWO
ROLL_BIT_0:
        rla     R5
        jc      ONE
        jmp     FIRST_DIGIT

EIGHT:
        add     #8, R11
        jmp     ROLL_BIT_2
FOUR:
        add     #4, R11
        jmp     ROLL_BIT_1
TWO:
        add     #2, R11
        jmp     ROLL_BIT_0
ONE:
        add     #1, R11

FIRST_DIGIT:
        add     #48, R11
        mov     R11, R4

        call    #ENTER_ASCII          ; Send ASCII Number to LCD Display

        mov     #0, R4                ; Clears R4
        mov     #0, R11               ; Clears R11
;-------------------------------------------------------------------------------
;-- THIS ROUTINE READS THE BCD NUMBER STORED IN R15 AND MAKES AN ASCII NUMBER --
;-------------------------------------------------------------------------------
        mov     #4, R10               ; Counter
        mov     #0, R7                ; Ensure R7 has 0 in it (Accumulator)
READ_4_BITS:
        rlc     R15
        jc      EIGHT_A
ROLL_BIT2:
        rlc     R15
        jc      FOUR_A
ROLL_BIT1:
        rlc     R15
        jc      TWO_A
ROLL_BIT0:
        rlc     R15
        jc      ONE_A
        jmp     DIGIT

EIGHT_A:
        add     #8, R7
        jmp     ROLL_BIT2
FOUR_A:
        add     #4, R7
        jmp     ROLL_BIT1
TWO_A:
        add     #2, R7
        jmp     ROLL_BIT0
ONE_A:
        add     #1, R7
;-------------------------------------------------------------------------------
;---- CREATE ASCII NUMBER ------------------------------------------------------
;-------------------------------------------------------------------------------
DIGIT:
        add     #48, R7               ; Make ASCII Number
        mov     R7, R4
```

```
        call     #ENTER_ASCII        ; Send ASCII Number to LCD

        mov      #0, R4              ; Clears R4
        mov      #0, R7              ; Clears R7
        dec      R10                 ; Decrement 1 from R10
        cmp      #0, R10             ; Does R10 = 0 ?
        jeq      NEW_NUMBER          ; If R10 = 0 go get new number
        jmp      READ_4_BITS         ; If R10 not 0 go read another 4 bits

NEW_NUMBER:
        mov      #0, &TA0CCR0        ; Clear out TA0CCR0

        mov.b    #00000000b, &BCSCTL1  ; Re-establish VLOCLK as CPU Clock
        mov.b    #00100000b, &BCSCTL3  ; Re-establish VLOCLK as CPU Clock
        mov.b    #11111110b, &BCSCTL2  ; Re-establish VLOCLK as CPU Clock

        reti                         ; Return from the Interupt
;-------------------------------------------------------------------------------
;----- SUBROUTINE THAT ENTERS ADC DATA (ASCII NUMBERS) INTO LCD DISPLAY ---------
;-------------------------------------------------------------------------------
ENTER_ASCII:
        mov.b    #00010000b, &P1OUT  ; (P1.4 = 1) LCD DISPLAY Data Mode Selected

        bis.b    #00100000b, &P1OUT  ; (P1.5) ENABLE pin brought HIGH

        mov      #26215, R6          ; 5 Millisecond Delay
DELAY_7:
        dec      R6
        jnz      DELAY_7

        mov.b    R4, &P2OUT          ; ASCII Numbers (Data) loaded into LCD Display

        bic.b    #00100000b, &P1OUT  ; P1.5 ENABLE pin brought LOW

        mov      #26215, R6          ; 5 Millisecond Delay
DELAY_8:
        dec      R6
        jnz      DELAY_8

        ret                          ; Returns program to after "call"
;-------------------------------------------------------------------------------
; Stack Pointer definition
;-------------------------------------------------------------------------------
        .global __STACK_END
        .sect   .stack
;-------------------------------------------------------------------------------
; Interrupt Vectors -- add in ".short" for the Interupt Service Routine
;-------------------------------------------------------------------------------
;       .sect   ".int00"    ; Slot 0
;       .sect   ".int01"    ; Slot 1
;       .sect   ".int02"    ; Slot 2    Port1_Vector
;       .sect   ".int03"    ; Slot 3    Port2_Vector
;       .sect   ".int04"    ; Slot 4
;       .sect   ".int05"    ; Slot 5    ADC10_Vector
;       .sect   ".int06"    ; Slot 6    USCIAB0TX_Vector
;       .sect   ".int07"    ; Slot 7    USCIAB0RX_Vector
;       .sect   ".int08"    ; Slot 8    TIMER0_A1_Vector
        .sect   ".int09"    ; TIMER_0 Vector
        .short  CAPTURE_ONE ; Interupt Service Routine
;       .sect   ".int10"    ; Slot 10   WDT_Vector
;       .sect   ".int11"    ; Slot 11   COMPARATORA_Vector
;       .sect   ".int12"    ; Slot 12   TIMER1_A1_Vector
;       .sect   ".int13"    ; Slot 13   TIMER1_A0_Vector
;       .sect   ".int14"    ; Slot 14   NMI_Vector
        .sect   ".reset"    ; MSP430 RESET Vector
        .short  RESET
;-------------------------------------------------------------------------
```

Fig. 7.7 Event capture

7.6   Up/Down Mode

The Up/Down mode counts up to a point then counts back down to zero (Fig. 7.8). A capture/compare (TACCRO) register is set where the timer starts counting down. Other capture/compare registers can be set at lower levels to generate different events, once on the way up and once on the way down. Refer to SLAU144K (www.ti.com) for details on how generating events is accomplished. Chapter. 8 explores using the timer in this mode to generate a PWM (pulse width modulation) signal.

7.7   Interrupts

Interrupts are just as the word implies. The microcontroller may be programmed to sit idle and wait for some event to happen, and when that event does happen it springs into action and go to a point in the code listing that accomplishes a task, then go back to sleep waiting

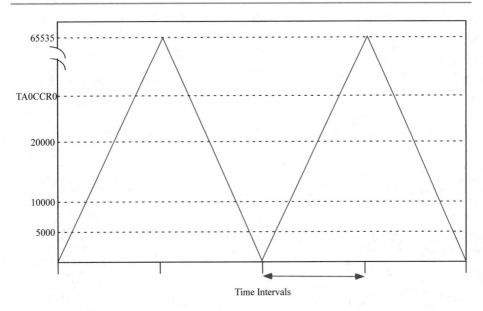

Fig. 7.8 Up/Down mode

for another event to happen. In essence, the microcontroller is in a "sleep state" and is interrupted.

Events that cause an interrupt can be set through the Timer_A module, through the I/O ports, as well as several of the other peripheral modules in the MSP430G2553. An important part of setting up an interrupt is the interrupt service routine vector table that is positioned at the bottom of code listings. Both of the code listings in this chapter have the full vector table included, however, most lines have been commented out because they are not used. Another important part of using interrupts is the "reti" statement that tells the microcontroller to go back to sleep after it has accomplished its tasks following response to the interrupt. In the TIMER_CAPTURE code listing, an interrupt is triggered by a pulse that is inputted into pin P1.1.

7.8   Post Lab Activities

1. In the Appendices we provide a brief introduction to Unified Modeling Language (UML) activity diagrams. These are UML compliant flow charts. Provide UML activity diagrams for programs provided in this laboratory exercise.
2. For assembly language programs provided; rewrite in C, compile, execute, and test for proper operation.

Lab 8: MSP430G2553 Timer_A Module (Pulse Width Modulation)

8

8.1 Purpose

In this lab the Timer_A module of the MSP430G2553 is configured to generate a Pulse Width Modulation (PWM) signal. PWM signals are used to control the speed of motors, control the brightness of lights, or vary voltage powering a device. PWM_OSCOPE demonstrates the generation of two PWM signals that are configured differently, and PWM_LED demonstrates the use of a PWM signal to control the brightness of a LED.

8.2 Discussion

Pulse Width Modulation (PWM) is a means of controlling an analog circuit with a digital signal. The digital signal is a square wave that is set at a specific frequency where the amount of time the signal is "High" and the amount of time the signal is "Low" is varied. Duty cycle is a term associated with PWM and refers to the percentage of the time the signal is "High."

In Fig. 8.1 the frequency and the period of both PWM signals are the same. The period of the waveform is 3.756 ms, and the frequency is 266 Hz. In Fig. 8.1 (top) one complete wave spends 50% of the time as "High" and 50% of the time as "Low," therefore, this PWM signal has a 50% duty cycle. In Fig. 8.1 (bottom) one complete wave spends 25% of the time as "High" and 75% of the time as "Low," and the duty cycle is 25%.

In this lab Timer_1 in the Timer_A module is configured in the UP/DOWN mode to generate the PWM signals. There are several "Output Modes" that can be selected through the Timer control registers, however, only two are used in this lab. Figure 8.2 depicts the PWM signals produced by Output Mode 2 and Output Mode 6. The Output Modes determine

© The Author(s), under exclusive license to Springer Nature Switzerland AG 2023   73
J. Kretzschmar et al., *MSP430 Microcontroller Lab Manual*, Synthesis Lectures on Digital
Circuits & Systems, https://doi.org/10.1007/978-3-031-26643-0_8

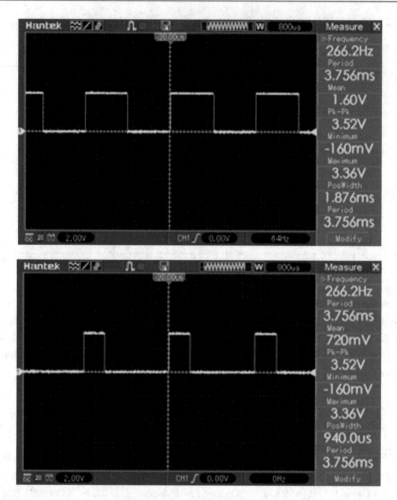

Fig. 8.1 Pulse width modulation. (top) 50% duty cycle, (bottom) 25% duty cycle

the state (High or Low) of the output signal. The state of the output signal is triggered when the timer count value is equal to the count values stored in the TA1CCR1 register. Refer to SLAU144K User Guide for details (www.ti.com).

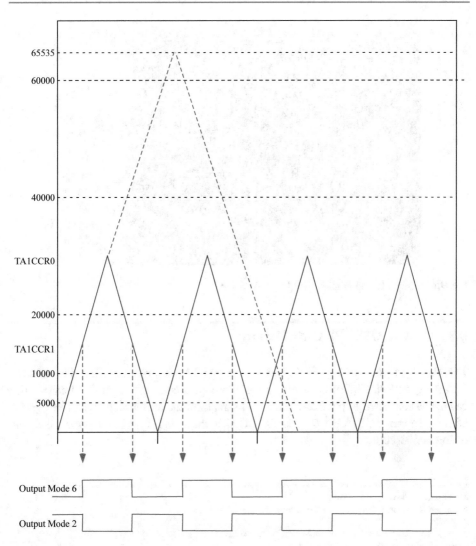

Fig. 8.2 Pulse width modulation Mode 2 and 6. Timer_1 configured in UP/DOWN mode. The vertical axis shows the count as the Timer progresses up and down. The horizontal axis is the timescale and is dependent upon the selected TAR clock speed. TA1CCR0 sets the top value of count. TA1CCR1 (and TA1CCR2 not shown) are selected to meet design specifications. This example demonstrates 50% duty cycle and two of seven possible output modes

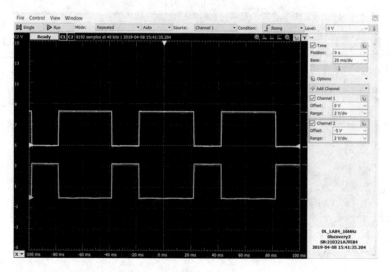

Fig. 8.3 Pulse width modulation. Output Mode 6 and 2

8.3   PWM_OSCOPE Code Listing

This code listing outputs two PWM signals, one on P2.1 (Output Mode 6), and the other on P2.4 (Output Mode 2). One channel of an oscilloscope is connected to pin P2.1 and another channel is connected to pin P2.4. The PWM signals on the oscilloscope look like Fig. 8.3. Change the values of TA1CCR1 and TA1CCR2 in the code listing and observe the signals on the oscilloscope.

```
;-----------------------------------------------------------------------
; Program Name: PWM_OSCOPE
; Assembly Code using Texas Instruments Code Composer
; Parts:
; -- Texas Instruments MSP-EXP430G2ET Launchpad with MSP430G2553
; -- Oscilloscope

; Documentation:
; -- Texas Instruments SLAU144K Users Guide (ti.com)
; -- Texas Instruments SLAS735J MSP430G2553 Datasheet (ti.com)
;
; Connections:
;
;    MSP430G2553              Oscilloscope
;        P2.1   ----------> Input (Channel 1)
;
;        P2.4   ----------> Input (Channel 2)
;        Ground              Ground
;
; This program sets up Timer_1 in the UP/DOWN mode with a CPU clock speed of 1 MHz
; and demonstrates the generation of two PWM signals from TIMER_1. The Timer is
```

```
; configured so that the Timer counts up to 30000 (set by TA1CCR0) and then counts
; down to 0. TA1CCR0 sets the value that the UP/DOWN Timer counts up to. The two
; PWM signals are outputted on P2.1 (TA1.1) and P2.4 (TA1.2). The output mode is
; selected by TA1CCTL1 and TA1CCTL2, and the set points for the output to occur
; are set by TA1CCR1 and TA1CCR2. Refer to the Users Guide SLAU144K for details.
; The output will occur on both the up-slope of the Timer and the down-slope of
; the Timer which will allow configuration of the PWM signal as needed. The
; different types of output (OUTMOD_x) are depicted in the Users Guide. With the
; TIMER_1 configured as described it will take 30 milliseconds to reach the top
; and then 30 milliseconds to get back to 0 (a period of 60 milliseconds). P2.1
; generates a PWM signal that is HIGH (outputting 3.3 volts) for 40 milliseconds
; and LOW (0 volts) for 20 milliseconds (duty cycle of 66.6%). P2.4 generates a
; PWM signal that is HIGH (outputting 3.3 volts) for 20 milliseconds and LOW
; (0 volts) for 40 milliseconds (duty cycle of 33.3%). Connect an oscilloscope to
; P2.1 and P2.4 to see the two PWM signals.
;
; J. Kretzschmar
;
;-------------------------------------------------------------------------------
            .cdecls C,LIST,"msp430.h"      ; Include device header file
            .def    RESET
            .text
            .retain
            .retainrefs
;-------------------------------------------------------------------------------
RESET       mov.w   #__STACK_END,SP        ; Initialize stackpointer
StopWDT     mov.w   #WDTPW|WDTHOLD,&WDTCTL  ; Stop watchdog timer
;-------------------------------------------------------------------------------
;    SETTING UP PARAMETERS FOR TIMER_1 ... ONCE CONFIGURED TIMER_1 WILL THEN
;            BE TURNED ON WITH "TA1CTL" (UP/DOWN MODE SELECTED ALSO)
;-------------------------------------------------------------------------------

     mov.b   &CALBC1_1MHZ, &BCSCTL1   ; Using 1 MHz Clock
     mov.b   &CALDCO_1MHZ, &DCOCTL    ; Using 1 MHz Clock

     mov.b   #00010010b, &P2SEL       ; P2.1 selected for (TA1.1) (CCI1A)
                                      ; P2.4 selected for (TA1.2) (CCI2A)

     mov.b   #00010010b, &P2DIR       ; P2.1 and P2.4 selected as outputs

     mov.w   #30000,  &TA1CCR0        ; TIMER_1 Capture/Compare Register_0
                                      ; Counts up to 30000 then down to 0
                                      ; Total for 1 UP/DOWN count = 60000
                                      ; Period = 60 milliseconds

     mov.w   #OUTMOD_6 | CCIS_0, &TA1CCTL1
                                      ; TIMER_1 Capture/Compare Control Register_1
                                      ; Bits 7-5 (110) selects Output Mode 6
                                      ; Bits 13-12 (00) selects CCI1A for (TA1.1)
                                      ; Bits 13-12 (01) selects CCI1B for (TA1.1)
                                      ; See SLAS735J for details
                                      ; See SLAU144JK for details
     mov.w   #10000,  &TA1CCR1        ; TIMER_1 Capture/Compare Register_1
                                      ; At 10000 Pin #9 (P2.1)(TA1.1) goes
                                      ; HIGH and stays HIGH until the timer
                                      ; goes below 10000, then it goes LOW

     mov.w   #OUTMOD_2 | CCIS_0, &TA1CCTL2
                                      ; TIMER_1 Capture/Compare Control Register_2
```

```
                                      ; Bits 7-5 (010) selects Output Mode 2
                                      ; Bits 13-12 (00) selects CCI1A for (TA1.2)
                                      ; Bits 13-12 (01) selects CCI1B for (TA1.2)
                                      ; See SLAS735J for details
                                      ; See SLAU144JK for details
         mov.w   #10000,  &TA1CCR2    ; TIMER_1 Capture/Compare Register_2
                                      ; At 10000 Pin 11 (P2.4)(TA1.2) goes
                                      ; LOW and stays LOW until the timer
                                      ; goes above 10000, then it goes HIGH

         mov.w   #TASSEL_2 | MC_3 | ID_0, &TA1CTL
                                      ; TIMER_1 Control Register
                                      ; Bits 9-8 (10) selects TASSEL_2 which
                                      ; selects SMCLK as the clock source
                                      ; Bits 5-4 (11) selects MC_3 which selects
                                      ; the UP/DOWN mode for the timer
                                      ; Bits 7-6 (00) selects ID_0 which selects
                                      ; SMCLK/1 (divide SMCLK by 1)
                                      ; See SLAU144JK for details
;-----------------------------------------------------------------------------
; Stack Pointer definition
;-----------------------------------------------------------------------------
         .global __STACK_END
         .sect   .stack
;-----------------------------------------------------------------------------
; Interrupt Vectors
;-----------------------------------------------------------------------------
         .sect   ".reset"            ; MSP430 RESET Vector
         .short  RESET
;-----------------------------------------------------------------------------
```

8.4   PWM_LED Code Listing

This code listing outputs a PWM signal on P2.1 which is connected to the red portion of the RGB LED. The jumper connector on the LaunchPad between P2.1 for the red portion of the RGB LED needs to be in place. The timer is configured for the Up/Down mode where the timer counts up to 30,000 and then counts back down to zero. The value of TA1CCR1 can be changed to produce a PWM signal with duty cycles of 12.5, 20, 25, 30, 50, and 75% by uncommenting the appropriate lines of code. You should be able to observe different brightness values of the LED as the duty cycle is changed.

```
;-----------------------------------------------------------------------

; Program Name: PWM_LED
; Assembly Code using Texas Instruments Code Composer
; Parts:
; -- Texas Instruments MSP-EXP430G2ET Launchpad with MSP430G2553
; -- Oscilloscope

; Documentation:
; -- Texas Instruments SLAU144K Users Guide (ti.com)
; -- Texas Instruments SLAS735J MSP430G2553 Datasheet (ti.com)
```

```
; This program sets up Timer_1 in the UP/DOWN mode with a CPU clock speed of
; 16 MHz and demonstrates the generation of a PWM signal on pin P2.1 that drives
; the brightness of the RED LED associated with P2.1. Pin P2.1 is part of the
; RGB LED consisting of P2.1 (RED), P2.3 (GREEN), and P2.5 (BLUE), however,
; each LED can be used individually by connecting the jumper and unconnecting
; the jumper for the other two LEDs. In this particular program only P2.1 can
; be used because it is designated as a peripheral use pin for the Timer. The
; value for the top of the UP/DOWN mode is selected and stored in TA1CCR0.
; Different values can be selected and stored in TA1CCR1 that will provide
; different PWM duty cycles which will change the brightness of the LED on P2.1.
; The PWM signal can be visualized by connecting an oscilloscope to pin P2.1.

; J. Kretzschmar

;-------------------------------------------------------------------------------
            .cdecls C,LIST,"msp430.h"      ; Include device header file
            .def    RESET
            .text
            .retain
            .retainrefs
;-------------------------------------------------------------------------------
RESET       mov.w   #__STACK_END,SP         ; Initialize stackpointer
StopWDT     mov.w   #WDTPW|WDTHOLD,&WDTCTL  ; Stop watchdog timer
;-------------------------------------------------------------------------------
    mov.b   &CALBC1_16MHZ, &BCSCTL1    ; Using 16 MHz Clock
    mov.b   &CALDCO_16MHZ, &DCOCTL     ; Using 16 MHz Clock
    mov.b   #DIVS_0, &BCSCTL2          ; MCLK/1 (divided by 1)

    mov.w   #30000, &TA1CCR0           ; TIMER_1 Capture/Compare Register_0
                                       ; Counts up to 30000 then down to 0
                                       ; Total for 1 UP/DOWN count = 60000
                                       ; Period = 3.75 milliseconds

    mov.w   #OUTMOD_2 | CCIS_0, &TA1CCTL1
                                       ; TIMER_1 Capture/Compare Control Register_1
                                       ; Bits 7-5 (110) selects Output Mode 2
                                       ; Bits 13-12 (00) selects CCI1A for (TA1.1)
                                       ; Bits 13-12 (01) selects CCI1B for (TA1.1)
                                       ; See SLAS735J for details
                                       ; See SLAU144K for details
;-------------------------------------------------------------------------------
;      UNCOMMENT THE VALUE YOU WANT FOR TA1CCR1
;-------------------------------------------------------------------------------

    mov.w   #3750,  &TA1CCR1     ; 12.5% Duty Cycle (2 x 3750 / 60000)
    ;mov.w  #6000,  &TA1CCR1     ; 20.0% Duty Cycle (2 x 6000 / 60000)
    ;mov.w  #7500,  &TA1CCR1     ; 25.0% Duty Cycle (2 x 7500 / 60000)
    ;mov.w  #9000,  &TA1CCR1     ; 30.0% Duty Cycle (2 x 9000 / 60000)
    ;mov.w  #15000, &TA1CCR1     ; 50.0% Duty Cycle (2 x 15000 / 60000)
    ;mov.w  #22500, &TA1CCR1     ; 75.0% Duty Cycle (2 x 22500 / 60000)


;-------------------------------------------------------------------------------

    bis.b   #00000010b, &P2DIR   ; P2.1 selected as output
    bis.b   #00000010b, &P2SEL   ; P2.1 selected for (TA1.1) (CCI1A)

    mov.w   #TASSEL_2 | MC_3 | ID_0, &TA1CTL
                                 ; TIMER_1 Control Register
```

```
                          ; Bits 9-8 (10) selects TASSEL_2 which
                          ; selects SMCLK as the clock source
                          ; Bits 5-4 (11) selects MC_3 which selects
                          ; the UP/DOWN mode for the timer
                          ; Bits7-6 (00) selects ID_0 which selects
                          ; SMCLK/1 (divide SMCLK by 1)
                          ; See SLAU144K for details
;-----------------------------------------------------------------------
; Stack Pointer definition
;-----------------------------------------------------------------------
          .global __STACK_END
          .sect    .stack
;-----------------------------------------------------------------------
; Interrupt Vectors
;-----------------------------------------------------------------------
          .sect   ".reset"              ; MSP430 RESET Vector
          .short  RESET
;-----------------------------------------------------------------------
```

8.5   Post Lab Activities

1. In the Appendices we provide a brief introduction to Unified Modeling Language (UML) activity diagrams. These are UML compliant flow charts. Provide UML activity diagrams for programs provided in this laboratory exercise.
2. For assembly language programs provided; rewrite in C, compile, execute, and test for proper operation.

Lab 9: MSP430G2553 UART Serial Communication

9.1 Purpose

The Universal Serial Communication Interface (USCI) peripheral on the MSP430G2553 supports the communication protocols of Serial Peripheral Interface (SPI), Inter–Integrated Circuit (I2C), and Universal Asynchronous Receiver/Transmitter (UART).

In this lab two MSP430G2553 microcontrollers are connected to demonstrate the UART mode function that is available in the MSP430G2553. One LaunchPad is configured as the transmitter, and the other LaunchPad is configured as the receiver. The transmitter generates ASCII characters, and the receiver displays the ASCII characters on the LCD display. This lab demonstrates the inner workings of packaged serial communication programs by looking at the movement of the actual "1s" and "0s" of data.

9.2 Discussion

UART is a serial communication protocol that can be one way (simplex) where the receiver has no provision to talk back to the transmitter, two-way (full duplex) where the transmitter and the receiver send and receive at the same time, or (half duplex) where the transmitter and receiver take turns sending and receiving.

When using the UART mode of serial communication, parameters for both the transmitter and receiver must be set up exactly the same. Parameters that can be selected when configuring the microcontrollers are the speed of transmission (Baud rate = bits per second), the number of bits to be sent in each package, whether a parity bit is included in the data package, and the number of stop bits to be included in the data package.

© The Author(s), under exclusive license to Springer Nature Switzerland AG 2023   81
J. Kretzschmar et al., *MSP430 Microcontroller Lab Manual*, Synthesis Lectures on Digital
Circuits & Systems, https://doi.org/10.1007/978-3-031-26643-0_9

In this lab the parameters are configured in the most basic form, simplex, 8 bits per data packet, no parity, and one stop bit. The sequence of events in the UART protocol consists of the following:

1. The data line is held HIGH until data is put on the line,
2. A start bit is added to the eight bits, the start bit pulls the data line LOW and signals to the receiver to get ready to receive eight bits of data,
3. The eight bits are sent out starting with the least significant bit (LSB) first,
4. One stop bit is added that pulls the data line HIGH signaling the end of the data package.

Sending the LSB first is the most common way of sending data, however, it is possible to send out the most significant bit (MSB) first if needed. At the receiver end the start bit and the stop bit are removed.

Refer to Fig. 9.1 to see how the UART "10 bit package" is put together. With a Baud rate of 9600 (9600 bits per second) and the CPU clock set at 1 MHz it takes 104 μs of time to send one bit. The small circle in the figure highlights the measurement information.

In Fig. 9.1, the large red circle highlights one 10–bit UART packet. From the Oscilloscope signal, you can see the pattern 0100011110. The first 0 is the start bit. The next seven bits are the data with the LSB of the data occurring first. The final 0 is inserted by the microcontroller to fill the eight data bit packet. Finally the signal ends up in a "High" state, waiting for another data packet transmission. This 10–bit packet is expanded in Fig. 9.2 and described in detail. To capture the oscilloscope screenshot the program was written to repeat sending ASCII number 113 with a short delay in between. Below are the parameters. Referencing SLAU144K (www.ti.com) the following parameters were selected:

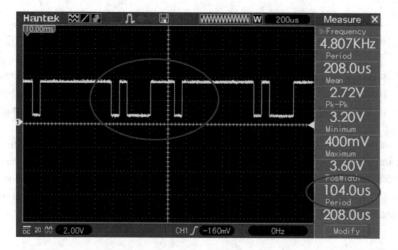

Fig. 9.1 A UART 10–bit packet

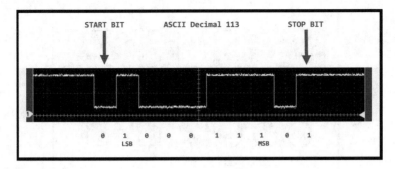

Fig. 9.2 Expanded UART 10–bit packet

1. CPU clock of 1 MHz,
2. Baud rate of 9600,
3. No parity bits,
4. 8–bit data package and LSB sent out first, and
5. 1 Start bit, 1 Stop bit.

9.3 Setting up the UART Demonstration

This demonstration requires two MSP430G2553 microcontrollers. One for Transmission and one for reception. For the Transmitter MSP430G2553, connect three tactile switches as shown in Fig. 9.3 (top). The tactile switch on pin P1.4 is pushed to input a "0," the tactile switch on pin P1.5 is pushed to input a "1," and the tactile switch on pin P1.3 is pushed to send out the 8–bit data packet.

Pin P1.2 on the Transmitter MSP430G2553 is the output for the UART signal and is connected to pin P1.1 of the MSP430G2553 that is the Receiver. There is a separate program for the Transmitter, and a separate program for the Receiver. Once the Transmitter program has been loaded into the Transmitter MSP430G2553 it can be powered by a 3 volt battery pack connected into one of the 3.3 volt pins on the LaunchPad.

When a "0" is entered the green LED on P2.3 flashes, when a "1" is entered the red LED on P2.1 flashes, and when 8 bits have been entered the blue LED on P2.5 stays on until the Send pushbutton has been pushed.

For the Receiver MSP430G2553, the LCD display is connected as has been done in the previous labs. Pin P1.1 receives the UART signal from the Transmitter MSP430G2553 and sends the received ASCII character to the LCD display.

See Fig. 9.3 as guides for configuring the Transmitter MSP430G2553 and the Receiver MSP430G2553. The code listing for the Transmitter is UART_TX, and the code listing for the Receiver is UART_RX. This is only a demonstration of how the UART module works in

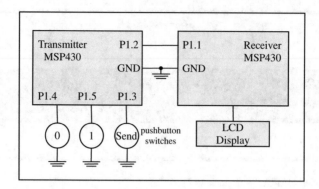

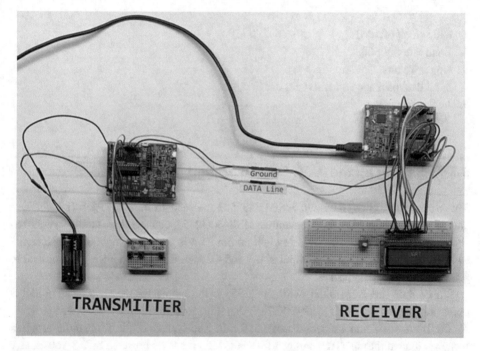

Fig. 9.3 UART demonstration

the MSP430G2553 and only one ASCII character can be sent at a time. Because of how the driver in the LCD display is configured for 40 characters on each line, an inputted message longer than 16 characters would need to have 24 "blank spaces" (ASCII 0100000) inputted before a character would appear on the second line. A possible further development of this lab would be to send the UART data wirelessly.

		Most significant digit							
		0x0_	0x1_	0x2_	0x3_	0x4_	0x5_	0x6_	0x7_
Least significant digit	0x_0	NUL	DLE	SP	0	@	P	`	p
	0x_1	SOH	DC1	!	1	A	Q	a	q
	0x_2	STX	DC2	"	2	B	R	b	r
	0x_3	ETX	DC3	#	3	C	S	c	s
	0x_4	EOT	DC4	$	4	D	T	d	t
	0x_5	ENQ	NAK	%	5	E	U	e	u
	0x_6	ACK	SYN	&	6	F	V	f	v
	0x_7	BEL	ETB	'	7	G	W	g	w
	0x_8	BS	CAN	(	8	H	X	h	x
	0x_9	HT	EM	)	9	I	Y	i	y
	0x_A	LF	SUB	*	:	J	Z	j	z
	0x_B	VT	ESC	+	;	K	[	k	{
	0x_C	FF	FS	'	<	L	\	l	|
	0x_D	CR	GS	-	=	M	]	m	}
	0x_E	SO	RS	.	>	N	^	n	~
	0x_F	SI	US	/	?	O	_	o	DEL

Fig. 9.4 ASCII table

To enter an ASCII character reference the ASCII table provided in Fig. 9.4. Since the 8–bit data packet input was selected, and ASCII characters are represented by a 7–bit binary number, a leading "0" needs to be inputted, then the 7–bit ASCII number starting from the most significant bit.

UART Transmitter

```
;-------------------------------------------------------------------------
; Program Name: UART_TRANSMITTER
; Assembly Code using Texas Instruments Code Composer
; Parts:
; -- Texas Instruments MSP-EXP430G2ET Launchpad with MSP430G2553
; -- 3 Tactile Switches
; -- 3.0 Volt Battery Pack (Connect to 3.3 volt pin and Ground on Launchpad)
; Documentation:
; -- Texas Instruments SLAU144K Users Guide (ti.com)
; -- Texas Instruments SLAS735J MSP430G2553 Datasheet (ti.com)
; Connections:
; -- Tactile Switch (Send button): one pole to P1.3 and one pole to Ground
; -- Tactile Switch ("0"): one pole to P1.4 and one pole to Ground
; -- Tactile Switch ("1"): one pole to P1.5 and one pole to Ground

; This program allows you to input data (ASCII characters) in binary via 3 push
; button switches (P1.4 enters a "0", P1.5 enters a "1", P1.3 sends the ASCII
; number). Push the "0" and "1" buttons quickly. There is only a very short
; delay built in to accomodate for switch bounce. Long pushes will cause multiple
; unintentional "0"s and "1"s to be inputted. The individual LEDs in the
; RGB LED are used as indicators (P2.1 RED = a "1" was entered)(P2.3 GREEN = a
; "0" was entered)(P2.5 BLUE = 8 Bits have been entered and are ready to send).
```

```
; ASCII characters are 7 bits long and this UART transmitter is configured for
; entry of an 8 bit packet, therefore, a "0" needs to be entered before the
; 7 bit ASCII number to account for 8 bits. Pushing the "SEND" button transmits
; the data to the RECEIVER going out on pin P1.2 labeled "TXD" on the LaunchPad.
; At this point the BLUE LED goes off and the TRANSMITTER is ready to have
; another 8 Bits of data inputted.
;
; J. Kretzschmar
;
;-------------------------------------------------------------------------------
            .cdecls C,LIST,"msp430.h"         ; Include device header file
            .def    RESET
            .text
            .retain
            .retainrefs
;-------------------------------------------------------------------------------
RESET       mov.w   #__STACK_END,SP           ; Initialize stackpointer
StopWDT     mov.w   #WDTPW|WDTHOLD,&WDTCTL    ; Stop watchdog timer
;-------------------------------------------------------------------------------
;                       SET UP CLOCK INPUTS AND OUTPUTS
;-------------------------------------------------------------------------------
    mov.b   &CALBC1_1MHZ, &BCSCTL1     ; Calibrated 1 MHz clock selected
    mov.b   &CALDCO_1MHZ, &DCOCTL      ; Calibrated 1 MHz clock selected

    mov.b   #00000100b, &P1DIR         ; P1.2 will be TX output
                                       ; Selected by P1SEL and P1SEL2
                                       ; P1.3, P1.4, P1.5 are pushbutton inputs
    mov.b   #00111000b, &P1REN         ; Enables Pull-up/Pull-down resistors
                                       ; on P1.3, P1.4 and P1.5
    mov.b   #00111000b, &P1OUT         ; Selects the Pull-up resistors for
                                       ; P1.3, P1.4 and P1.5

    mov.b   #11111111b, &P2DIR         ; All pins are outputs
                                       ; Using the RGB LED
                                       ; P2.3 (GREEN) a "0" was entered
                                       ; P2.1 (RED) a "1" was entered
                                       ; P2.5 (BLUE) 8 Bits have been entered and
                                       ; are ready to "SEND"
    bic.b   #00101010b, &P2OUT         ; Makes sure LEDs are off
;-------------------------------------------------------------------------------
;                       SET UP UART TRANSMITTER
;-------------------------------------------------------------------------------
    mov.b   #00000110b, &P1SEL         ; UART RX and TX pins
    mov.b   #00000110b, &P1SEL2        ; UART RX and TX pins
;-------------------------------------------------------------------------------
;                   Register UCA0CTL0 is all "0"s
;   (BIT 7 = 0 Parity disabled)(BIT 6 = 0 Do not need to select a parity)
;   (BIT 5 = 0 LSB sent out first)(BIT 4 = 0 8 Bit data mode)
;   (BIT 3 = 0 One stop Bit)(BIT 1&2 = 0 UART Mode)(Bit 0 = 0 Asynchronous mode)
;-------------------------------------------------------------------------------

    bis.b   #UCSWRST, &UCA0CTL1        ; Reset UART ... ON
                                       ; Reset has to be ON to set parameters
    mov.b   #UCSSEL_2, &UCA0CTL1       ; SMCLK selected as the clock source
    mov.b   #104, &UCA0BR0             ; 9600 Baud selected
                                       ; (SLAU144K Table 15-4, p.448)
    mov.b   #0,    &UCA0BR1            ; 9600 Baud rate selected
    mov.b   #UCBRS0, &UCA0MCTL
```

```
        bic.b    #UCSWRST, &UCA0CTL1        ; Reset  UART ... OFF
                                           ; Reset has to be OFF for UART to operate
;---------------------------------------------------------------------------------
;    THIS SECTION INPUTS 8 BITS OF DATA INTO R10 BY ENTERING "1"s
AND "0"s ;    BY PUSHBUTTONS ON P1.4 AND P1.5   EACH TIME A BUTTON
IS PUSHED A LED ;    WILL COME ON FOR 65 MILLISECONDS INDICATING A
BUTTON WAS PUSHED ... THEN ;    THE LED WILL GO OFF FOR 65
MILLISECONDS BEFORE THE PROGRAM CONTINUES. WHEN ;    8 BITS HAVE
BEEN ENTERED THE BLUE LED COMES ON AND THE 8 BITS ARE READY TO ; BE
SENT OUT ON P1.2
;---------------------------------------------------------------------------------
;               THE INPUTTED "1"s AND "0"s ARE STORED IN R10
;---------------------------------------------------------------------------------
RELOAD_R10:
        mov      #8, R5                    ; Move the number 8 into the counter
TEST:
        cmp      #0, R5                    ; Is the counter at zero?
        jeq      EIGHT_BITS_IN_R10         ; If 0 goto EIGHT_BITS_IN R10
        bit.b    #00010000b, &P1IN         ; Has the "0" button been pushed?
        jz       GATE_0                    ; If so goto GATE_0
        bit.b    #00100000b, &P1IN         ; Has the "1" button been pushed?
        jz       GATE_1                    ; If so goto GATE_1
        jmp      TEST                      ; Goto TEST and keep checking the buttons

GATE_0:
        add      R10, R10                  ; A "0" gets pushed into R10
        dec      R5                        ; Counter is decremented by 1
        bis.b    #00001000b, &P2OUT        ; GREEN LED goes ON
        mov      #65535, R8                ; Approx 65 milliseconds
DELAY_ZERO_ON:                            ; DELAY Routine (ON)
        dec      R8
        jnz      DELAY_ZERO_ON
        bic.b    #00001000b, &P2OUT        ; GREEN LED goes OFF
        mov      #65535, R8                ; Approx 65 milliseconds
DELAY_ZERO_OFF:                           ; DELAY Routine (OFF)
        dec      R8
        jnz      DELAY_ZERO_OFF
        jmp      TEST                      ; Go back to testing the buttons

GATE_1:
        add      R10, R10                  ; A "0" gets pushed into R10
        bis      #1,  R10                  ; A "1" gets added to R10
        dec      R5                        ; Counter is decremented by 1
        bis.b    #00000010b, &P2OUT        ; RED LED goes ON
        mov      #65535, R8                ; Approx 65 milliseconds
DELAY_ONE_ON:                             ; DELAY Routine (ON)
        dec      R8
        jnz      DELAY_ONE_ON
        bic.b    #00000010b, &P2OUT        ; RED LED goes OFF
        mov      #65535, R8                ; Approx 65 milliseconds
DELAY_ONE_OFF:                            ; DELAY Routine (OFF)
        dec      R8
        jnz      DELAY_ONE_OFF
        jmp      TEST                      ; Go back to testing the buttons

EIGHT_BITS_IN_R10:
        bis.b    #00100000b, &P2OUT        ; BLUE LED "ON" indicates data is ready to SEND
WAITING_TO_SEND:                          ; Testing if the "SEND" has been pushed
        bit.b    #00001000b, &P1IN         ; Has "SEND" button been pushed
```

```
       jnz     WAITING_TO_SEND       ; Keep checking the "SEND" button
;------------------------------------------------------------------------------
;   IF THE "SEND" BUTTON WAS PUSHED THE PROGRAM CONTINUES ON
;------------------------------------------------------------------------------
       bic.b   #00100000b, &P2OUT    ; BLUE LED goes "OFF"
;------------------------------------------------------------------------------
;    IFG2 is the Interrupt Flag Register 2 and controls the flow of data
;    within the UART module for both the Transmitter and the Receiver. On the
;    Transmitter side if the UCA0TXIFG Bit is not set (i.e. = 0) then the transmit
;    buffer still has data in it and is not ready to accept new data. If UCA0TXIFG
;    is set (i.e. = 1) then the transmit buffer is empty and ready to accept new
;    data to be sent out on P1.2.
;------------------------------------------------------------------------------
CHECK_STATUS_OF_TRANSMIT_BUFFER:
       bit.b   #UCA0TXIFG, &IFG2                    ; Test the UCA0TXIFG flag
       jz      CHECK_STATUS_OF_TRANSMIT_BUFFER ; If "0" keep checking
;------------------------------------------------------------------------------
;                        IF "1" THE PROGRAM CONTINUES ON
;------------------------------------------------------------------------------
       mov     R10, &UCA0TXBUF   ; R10 moved into the Transmit Buffer and sent out

       jmp     RELOAD_R10        ; Goto the top and reload the next 8 Bits into R10
;------------------------------------------------------------------------------
; Stack Pointer definition
;------------------------------------------------------------------------------
           .global __STACK_END
           .sect   .stack
;------------------------------------------------------------------------------
; Interrupt Vectors
;------------------------------------------------------------------------------
           .sect   ".reset"               ; MSP430 RESET Vector
           .short  RESET
                                                   .
;------------------------------------------------------------------------------
```

UART Receiver

```
;------------------------------------------------------------------------------
; Program Name: UART_RCEIVER
; Assembly Code using Texas Instruments Code Composer
; Parts:
; -- Texas Instruments MSP-EXP430G2ET Launchpad with MSP430G2553
; -- Newhaven NHD-0216K1Z 16x2 LCD Display (or similar display)
; -- 10K Potentiometer (for contrast control)
; Documentation:
; -- Texas Instruments SLAU144K Users Guide (ti.com)
; -- Texas Instruments SLAS735J MSP430G2553 Datasheet (ti.com)
; -- Newhaven NHD-0216K1Z Datasheet (newhavendisplay.com)
; -- Paper "Controlling a 16x2 LCD Display with the MSP430G2553 Microcontroller
;    Using Assembly Code" (Appendix)
;
; This program receives data (ASCII characters) through P1.1 labeled "RXD" on
; the LaunchPad and displays the ASCII character on the LCD display. The LCD
; program is the same that has been used before and the timing parameters are
; established using the 16 MHz calibrated clock. The UART Receiver needs the
; 1 MHz calibrated clock for establishing the timing parameters in the UART
; module ... so the clock is changed back and forth as needed.
;
```

```
; Connections:
;
;  LCD Display      MSP430G2553
;    Pin #1           Ground
;    Pin #2           +5 volts
;    Pin #3           Contrast control (center pin from 10K potentiometer)
;    Pin #4           P1.4 (R/S Register Select)
;    Pin #5           Ground (Read/Write selection: 0 for Write, 1 for Read)
;    Pin #6           P1.5 (Enable)
;    Pin #7           P2.0 (Data)
;    Pin #8           P2.1 (Data)
;    Pin #9           P2.2 (Data)
;    Pin #10          P2.3 (Data)
;    Pin #11          P2.4 (Data)
;    Pin #12          P2.5 (Data)
;    Pin #13          P2.6 (Data)
;    Pin #14          P2.7 (Data)
;    Pin #15          +5 volts (backlight)
;    Pin #16          Ground (backlight)
;
; J. Kretzschmar


;-------------------------------------------------------------------------------
            .cdecls C,LIST,"msp430.h"      ; Include device header file
            .def    RESET
            .text
            .retain
            .retainrefs
;-------------------------------------------------------------------------------
RESET       mov.w   #__STACK_END,SP        ; Initialize stackpointer
StopWDT     mov.w   #WDTPW|WDTHOLD,&WDTCTL  ; Stop watchdog timer
;-------------------------------------------------------------------------------
;----- INITIAL SETUP MSP430G2553 PARAMETERS ------------------------------------
;-------------------------------------------------------------------------------
START_HERE:
    mov.b   &CALBC1_16MHZ,&BCSCTL1  ; Calibrated 16 MHz CPU Clock
    mov.b   &CALDCO_16MHZ,&DCOCTL   ; Calibrated 16 MHz CPU Clock

    mov.b   #00000000b, &P1SEL      ; Port 1 selected as I/O pins
    mov.b   #00000000b, &P2SEL      ; Port 2 selected as I/O pins
    mov.b   #00110000b, &P1DIR      ; P1.4 and P1.5 designated as OUTPUT pins
    mov.b   #11111111b, &P2DIR      ; P2.0 - P.7 designated as OUTPUT pins
    mov.b   #00000000b, &P1OUT      ; All Port_1 pins are cleared
    mov.b   #00000000b, &P2OUT      ; All Port_2 pins are cleared
;-------------------------------------------------------------------------------
;----- WAKE UP SEQUENCE OF LCD DISPLAY --- INITIAL 50 MILLISECOND DELAY ---------
;-------------------------------------------------------------------------------
    mov     #65535, R6         ; 12.5 Millisecond Delay
DELAY_0:
    dec     R6
    jnz     DELAY_0
    mov     #65535, R6         ; 12.5 Millisecond Delay
DELAY_00:
    dec     R6
    jnz     DELAY_00
    mov     #65535, R6         ; 12.5 Millisecond Delay
DELAY_000:
    dec     R6
    jnz     DELAY_000
```

```
    mov      #65535, R6           ; 12.5 Millisecond Delay
DELAY_0000:
    dec      R6
    jnz      DELAY_0000
;----- (COMMAND: FIRST WAKE UP CALL --- SELECTS 8-BIT, 2 LINE) -----------------
    mov.b    #00111000b, R4       ; 8-Bit entry format, 2 Line
    call     #SETUP
;----- (COMMAND: SECOND WAKE UP CALL --- SELECTS 8-BIT, 2 LINE) ----------------
    mov.b    #00111000b, R4       ; 8-Bit entry format, 2 Line
    call     #SETUP
;----- (COMMAND: THIRD WAKE UP CALL --- SELECTS 8-BIT, 2 LINE) -----------------
    mov.b    #00111000b, R4       ; 8-Bit entry format, 2 Line
    call     #SETUP
;------------------------------------------------------------------------------
;----- SETUP INFORMATION TO BE ENTERED INTO THE LCD DISPLAY --------------------
;------------------------------------------------------------------------------
;----- (COMMAND: CLEAR DISPLAY) -----------------------------------------------
    mov.b    #00000001b, R4       ; Clear Display
    call     #SETUP
;----- (COMMAND: SETS DDRAM ADDRESS --- WHERE CHARACTERS APPEAR) ---------------
    mov.b    #10000000b, R4       ; DDRAM Address 00 selected
    call     #SETUP
;----- (COMMAND: DISPLAY CONTROL) ---------------------------------------------
    mov.b    #00001100b, R4       ; Display ON, Cursor OFF, Cursor Blinking OFF
    call     #SETUP
;----- (COMMAND: DISPLAY SHIFT/CURSOR MOVE (RIGHT OR LEFT) --------------------
    mov.b    #00000110b, R4       ; Display Shift OFF, Next character to the right
    call     #SETUP
;----- (COMMAND: CURSOR/DISPLAY SHIFT)(NOT IMPLEMENTED) -----------------------
    ;mov.b   #00010000b, R4       ; Cursor/Display Shift OFF, Shift Left
    ;call    #SETUP
;----- (COMMAND: CURSOR HOME)(NOT IMPLEMENTED) --------------------------------
    ;mov.b   #00000010b, R4       ; Cursor Home (DDRAM address 00)
    ;call    #SETUP
    jmp      RECEIVER       ; All done with setup, goto RECEIVER
;------------------------------------------------------------------------------
;----- SUBROUTINE THAT ENTERS SETUP INFORMATION INTO LCD DISPLAY ---------------
;------------------------------------------------------------------------------
SETUP:
    mov.b    #00000000b, &P1OUT   ; (P1.4 = 0) LCD DISPLAY Command Mode Selected

    bis.b    #00100000b, &P1OUT   ; (P1.5) ENABLE pin brought HIGH

    mov      #26215, R6           ; 5 Millisecond Delay
DELAY_1:
    dec      R6
    jnz      DELAY_1

    mov.b    R4, &P2OUT           ; Setup Information loaded into LCD DISPLAY

    bic.b    #00100000b, &P1OUT   ; (P1.5) ENABLE pin brought LOW

    mov      #26215, R6           ; 5 Millisecond Delay
DELAY_2:
    dec      R6
    jnz      DELAY_2

    ret                          ; Returns program to after "call"
;------------------------------------------------------------------------------
```

```
;                        SET UP UART RECEIVER
;-------------------------------------------------------------------------------
RECEIVER:
      mov.b    &CALBC1_1MHZ, &BCSCTL1 ; Calibrated 1 MHZ CPU Clock
      mov.b    &CALDCO_1MHZ, &DCOCTL  ; Calibrated 1 MHZ CPU Clock
;-------------------------------------------------------------------------------
;                      Register UCA0CTL0 is all "0"s
;    (BIT 7 = 0 Parity disabled)(BIT 6 = 0 Do not need to select a parity)
;    (BIT 5 = 0 LSB sent out first)(BIT 4 = 0   8 Bit data mode)
;    (BIT 3 = 0 One stop Bit)(BIT 1&2 = 0 UART Mode)(Bit 0 = 0 Asynchronous mode)
;-------------------------------------------------------------------------------

      mov.b    #00000110b, &P1SEL     ; UART RX and TX pins
      mov.b    #00000110b, &P1SEL2    ; UART RX and TX pins

      bis.b    #UCSWRST, &UCA0CTL1    ; Reset UART ... ON
                                      ; Reset has to be ON to configure parameters
      mov.b    #UCSSEL_2, &UCA0CTL1   ; SMCLK selected as the clock source
      mov.b    #104, &UCA0BR0         ; 9600 Baud (SLAU144K Table 15-4, p.448)
      mov.b    #0,   &UCA0BR1         ; 9600 Baud rate selected
      mov.b    #UCBRS0, &UCA0MCTL

      bic.b    #UCSWRST, &UCA0CTL1    ; Reset UART ... OFF
                                      ; Reset has to be OFF for UART to operate
;-------------------------------------------------------------------------------
;    IFG2 is the Interrupt Flag Register 2 and controls the flow of data
;    within the UART module for both the Transmitter and the Receiver. On the
;    Receiver side if the UCA0RXIFG Bit is not set (i.e. = 0) then the receive
;    buffer has not received data yet from the transmitter. If UCA0RXIFG is set
;    (i.e. = 1) then the receive buffer has received a complete 8 Bits of data and
;    is ready to move that data through the UART module to be displayed on the
;    LCD display.
;-------------------------------------------------------------------------------
CHECK_STATUS_OF_RECEIVE_BUFFER:
      bit.b    #UCA0RXIFG, &IFG2              ; Test the UCA0RXIFG flag
      jz       CHECK_STATUS_OF_RECEIVE_BUFFER ; If "0" keep checking
;-------------------------------------------------------------------------------
;            IF "1" THE PROGRAM CONTINUES ON
;-------------------------------------------------------------------------------

      mov.b    &UCA0RXBUF, R7         ; Contents of Receive buffer moved into R7

      mov      R7, R4                 ; R7 moved into R4
      call     #ENTER_ASCII           ; Send to LCD display

      jmp      CHECK_STATUS_OF_RECEIVE_BUFFER  ; Go check for another 8 Bits of data

;-------------------------------------------------------------------------------
;----- SUBROUTINE THAT ENTERS DATA (ASCII NUMBERS) INTO LCD DISPLAY -------------
;-------------------------------------------------------------------------------
ENTER_ASCII:

      mov.b    &CALBC1_16MHZ, &BCSCTL1 ; Calibrated 16 MHz CPU Clock
      mov.b    &CALDCO_16MHZ, &DCOCTL  ; Calibrated 16 MHz CPU Clock

      mov.b    #00010000b, &P1OUT  ; (P1.4 = 1) LCD DISPLAY Data Mode Selected

      bis.b    #00100000b, &P1OUT  ; (P1.5) ENABLE pin brought HIGH
```

```
        mov     #26215, R6           ; 5 Millisecond Delay
DELAY_3:
        dec     R6
        jnz     DELAY_3

        mov.b   R4, &P2OUT           ; ASCII Numbers (Data) loaded into LCD Display

        bic.b   #00100000b, &P1OUT   ; P1.5 ENABLE pin brought LOW

        mov     #26215, R6           ; 5 Millisecond Delay
DELAY_4:
        dec     R6
        jnz     DELAY_4

        mov.b   &CALBC1_1MHZ, &BCSCTL1; Clock reset for UART

        mov.b   &CALDCO_1MHZ, &DCOCTL; Clock reset for UART

        ret                          ; Returns program to after "call"
;-----------------------------------------------------------------------
; Stack Pointer definition
;-----------------------------------------------------------------------
        .global __STACK_END
        .sect   .stack
;-----------------------------------------------------------------------
; Interrupt Vectors
;-----------------------------------------------------------------------
        .sect   ".reset"                 ; MSP430 RESET Vector
        .short  RESET
;-----------------------------------------------------------------------
```

9.4   Post Lab Activities

1. In the Appendices we provide a brief introduction to Unified Modeling Language (UML) activity diagrams. These are UML compliant flow charts. Provide UML activity diagrams for programs provided in this laboratory exercise.
2. For assembly language programs provided; rewrite in C, compile, execute, and test for proper operation.

Lab 10: MSP430G2553 Applications

10

10.1 Introduction

These projects have been included to demonstrate practical applications for using the MSP430G2553 microcontroller. The projects are not intended to be built exactly as described, however, they are presented to show capabilities of the microcontroller that may be incorporated into a particular application. Schematic diagrams, code listings, and circuit board drawings are provided. The code listings are well documented and provide information about pin inputs and outputs.

In these examples the circuit boards were prepared by using a pencil to draw the lines on the copper side of the board and then cutting the lines with a small cutting disk mounted on a rotary tool. This process is best done outside and following safety precautions of wearing a mask and using protective eyewear. The copper can then be cleaned with a steel wool soap pad.

Additionally, an oscilloscope or Volt/Ohm meter may be used to test the circuits to verify that they are working. There are many varieties of these test instruments, therefore, only generic instructions will be provided when recommending verification.

The chapter provides details for completing projects using skills developed throughout the lab manual. These projects include:

- External clock sources,
- Twelve button keypad matrix,
- Reading the contents of a register, and
- Reading an LM34 temperature sensor.

© The Author(s), under exclusive license to Springer Nature Switzerland AG 2023
J. Kretzschmar et al., *MSP430 Microcontroller Lab Manual*, Synthesis Lectures on Digital
Circuits & Systems, https://doi.org/10.1007/978-3-031-26643-0_10

10.2   Project 1: External Clock Sources

Many microcontrollers have timers as peripherals that function based on CPU clock. Although the internal CPU clock is fairly accurate, sometimes you need very precise timing which can be realized with an external crystal oscillator. The MSP430 Launchpad has a 32.768 KHz crystal/oscillator external to the microcontroller chip on the board, however, to engage that oscillator to function as the CPU clock necessitates designating pins number 18 and 19 on the microcontroller that could otherwise be used for another purpose.

In addition, two small surface mount resistors would need to be moved which is a difficult process. This external oscillator project allows you to input an external signal for use as the CPU clock to attain precise timing signals when using the timer module.

10.2.1 32.768 KHz Crystal Oscillator

The oscillator is constructed using one section of a 4069 Hex Inverter IC, a 32.768 KHz crystal, two resistors, and two capacitors. The circuit is powered by the $+5$ V available onboard the Launchpad. For ease of construction, the 14 pin IC socket for the 4069 (Hex Inverter IC) is modified by removing pins 2, 5, 6, 9, 10, 11, 12, and 13 from the socket that are not used. Although pins #1 and #8 are not electrically connected in the circuit they are left in place to help anchor the socket. To remove the unused pins, push the pin up from the bottom with a needle nose pliers and then pull it out from the top.

Project details are provided in Fig. 10.1 and a picture of the project is shown in Fig. 10.3.

10.2.2 Clock Oscillator

Clock oscillators are used in the electronics industry to establish a reference frequency used for timing purposes. These oscillators are made in many different packages and frequencies. The package style used in this project is a 4 pin "can" style that fits a 14 pin IC socket. The 5 V to power the oscillator is supplied from the Launchpad to Pin #14, Pin #7 is the ground connection, Pin #8 is the output, and Pin #1 is not connected.

The external clock signal can be inputted into microcontroller pin P1.0 configured as a peripheral use pin through register P1SEL then selected as TA0CLK input (clock for timer A0) in the TA0CTL register. Reference Fig. 10.2.

```
;-----------------------------------------------------------------------------
; Program Name: EXTERNAL_CLOCK
; Assembly Code using Texas Instruments Code Composer
;
; Parts:
; -- Texas Instruments MSP-EXP430G2ET Launchpad with MSP430G2553
; -- External clock source as described in text
;
; Documentation:
; -- Texas Instruments SLAU144K Users Guide (ti.com)
; -- Texas Instruments SLAS735J MSP430G2553 Datasheet (ti.com)
;
; This program uses an external clock source for the Timer function on the TI
```

```
; MSP430G2553 microcontroller. This provides the capability to use a more
; accurate clock source (a crystal oscillator) than the onboard DCO clock
; for timing applications. The frequencies of 1, 4, 5, 10, 12, 20, and 24 MHz
; were used with success. The 32.768 KHz external crystal oscillator was also
; tested with success. An on/off pulse was generated on Pin P1.6 that was
; observed on an oscilloscope. Depending on the 3 variables of (1) External clock
; frequency, (2) Division of clock frequency by 1, 2, 4, or 8, and (3) What
; TA0CCR0 is set at ... the frequency of the pulse produced at P1.6 can be
; designed for your specific application.
;
; J. Kretzschmar
;
;-------------------------------------------------------------------------------
            .cdecls C,LIST,"msp430.h"       ; Setup Information
            .def    RESET
            .text
            .retain
            .retainrefs
;-------------------------------------------------------------------------------
RESET       mov.w   #__STACK_END,SP         ; Initialize stackpointer
StopWDT     mov.w   #WDTPW|WDTHOLD,&WDTCTL   ; Stop watchdog timer

            mov.b   #00000001b, &P1SEL      ; P1.0 selected as peripheral use pin
                                            ; (This is the external clock input)
            mov.b   #01000000b, &P1DIR      ; P1.6 selected as output pin
            bis.b   #01000000b, &P1OUT      ; P1.6 is turned ON initially

            mov.w   #CCIE, &TA0CCTL0        ; Capture/Compare interrupt enabled
            mov.w   #TASSEL_0 | MC_1 | ID_0 | TACLR, &TA0CTL
                                            ; TASSEL_0 = External clock selection
                                            ; MC_1 = Up Mode (counts to TA0CCR0)
                                            ; ID_0 = Clock Frequency/1
                                            ; TACLR = Timer cleared (starts at 0)
            mov.w   #16384, &TA0CCR0        ; Counts up to 16348 which triggers an
                                            ; interrupt
;---------------- Continually loops looking for an Interrupt ------------------
LOOP:
            bis.w   #GIE, SR                ; Global interrupts enabled
            jmp     LOOP                    ; Keeps looking for an interrupt
;---------------- Timer Interrupt Routine -------------------------------------
ONOFF:
            xor     #01000000b, &P1OUT      ; XORs P1.6 (creates On/Off event)
            reti                            ; Returns from interrupt ... goes back
                                            ; to the LOOP and looks for another
                                            ; interrupt
;-------------------------------------------------------------------------------
; Stack Pointer definition
;-------------------------------------------------------------------------------
            .global __STACK_END
            .sect   .stack
;-------------------------------------------------------------------------------
; Interrupt Vectors
;-------------------------------------------------------------------------------
            .sect   ".reset"                ; MSP430 RESET Vector
            .short  RESET
            .sect   ".int09"                ; Timer 0 interrupt
            .short  ONOFF                   ; Goes to ONOFF

;-------------------------------------------------------------------------------
```

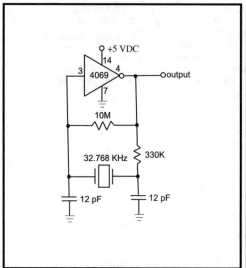

a) 32.768 KHz Crystal Oscillator.

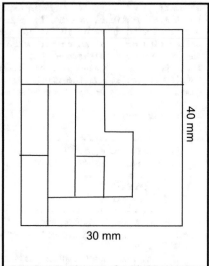

b) 32.768 KHz Oscillator Circuit Board
Layout (Actual Size).

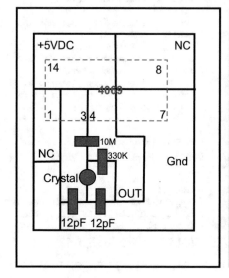

c) Circuit board parts placement.

d) 32.768 KHz oscillator.

Fig. 10.1 External clock source with a 32.768 KHz oscillator

a) Clock oscillators.

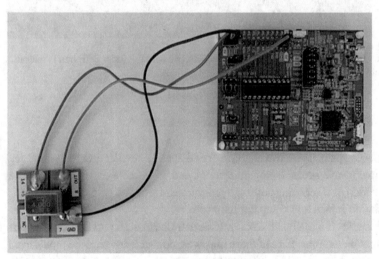

b) Clock oscillator output into MSP430 pin 1.0.

Fig. 10.2 External clock oscillator

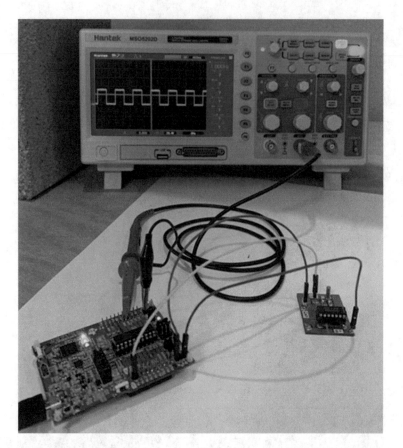

Fig. 10.3 External 32.768 KHz crystal oscillator used with the timer peripheral generating a precise 1 Hz signal

10.3 Project 2: 12 Button Keypad Matrix

Pushbutton keypads are used in many devices that we encounter in day–to–day life. Examples include microwave ovens, laptop computers, and TV remote controllers. Most keypads use a combination of a scanning software program and a matrix of tactile switches to decrease the number of inputs into computer hardware.

For example, the MSP430G2553 microcontroller has 16 pins that can be configured for use as input/output pins. If 12 of those pins were connected to read the inputs from 12 tactile switches then there would be only 4 pins remaining to use for other purposes. With the use of a matrix of switches and scanning software those same 12 switches can be read using 7 input/output pins on the microcontroller. This project will demonstrate how this can be accomplished.

10.3.1 How the Keypad Matrix Works

The 12 tactile switches are arranged in a pattern of 3 columns and 4 rows. Each switch has an associated diode to ensure that when that particular switch is pushed it is the only one detected by the microcontroller (Fig. 10.4).

The MSP430G2553 microcontroller is programmed to sense which of the 12 tactile switches has been pushed. The CPU clock is set to 1 MHz divided by 8. This means that the CPU clock is operating at 125 KHz and one cycle time takes 8 μs.

The assembly program that accompanies this section operates in the following manner:

1. Pins P1.0, P1.1, P1.2, and P1.3 are designated as output pins, and Pins P1.5, P1.6, and P1.7 are designated as input pins. Pins P1.0, P1.1, P1.2, and P1.3 are the rows, and P1.5, P1.6, and P1.7 are the columns.
2. Sequentially each row is brought high (3.3 V placed on the output Pins P1.0, P1.1, P1.2, and P1.3) and when the row is high then each column (input pins P1.5, P1.6, and P1.7) is tested one at a time to see if that pin has 3.3 V on it. This would signify that the tactile switch located at that particular row and column position was pushed.
3. If one of the column input pins becomes high (has 3.3 V on it) then the software continues to test the column as to when it becomes low. This signifies that the tactile switch has been released.
4. The program continues on to process the data entered through the 12 tactile switches.

For the program to complete one scan of the 12 switches takes approximately 1000 μs (1.0 ms). Although a tactile switch may seem to operate very fast, even a quick push and release takes about 200 ms. Considering how fast the microcontroller is scanning the switches, without some sort of "special programming" in the code the program may sense that a switch has been pushed 200 times when in reality it was only pushed once.

In the oscilloscope/logic analyzer screenshot shown in Fig. 10.5 the four green traces at the bottom show the rows (P1.0, P1.1, P1.2, and P1.3) being brought high (3.3 V), and the blue trace shows when the switch was pushed. The quick push of a switch (blue trace) took 144 ms. During that period of time the microcontroller sensed that the switch had been pushed many times (approximately 250 pulses on the blue line).

To avoid the microcontroller sensing that a switch has been pushed multiple times instead of just once, special code is incorporated into the program that will sense when the switch is released before advancing on to execute what will happen when the switch is pushed (one time). The first sample code is for sensing when the switch has been pushed. The second sample code is for sensing when the switch has been released and avoids sensing that the tactile switch has been pushed multiple times.

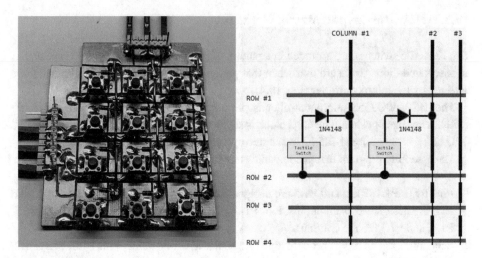

Fig. 10.4 12 tactile switches and matrix diagram (For clarity only two tactile switches are shown in the diagram. Solid black circles indicate connections)

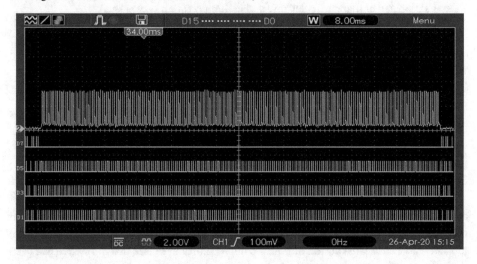

Fig. 10.5 12 button keypad switch bounce

10.3.2 Sample Code: Reading a Positive Voltage at P1.4 (Switch Pushed)

This code snippet tests for a logic high voltage on a pin. Input connections are subject to picking up stray signals from radio interference or static that may trigger an unintended event. To minimize this possibility the input pin is connected through a relatively low value resistor (10,000 Ω) to ground. This is called a Pull–down resistor and code can be written to engage the built–in Pull–down (or Pull-up) resistors in MSP430 microcontrollers. The code

```
mov.b      #00000000b, &P1SEL    ; Selects pins as GPIO and clears P1 Port
mov.b      #01000000b, &P1DIR    ; Selects pin P1.6 as OUTPUT, other pins are INPUT
mov.b      #10111111b, &P1REN    ; Enables Pull-up/Pull-down resistors on INPUT pins
mov.b      #10101111b, &P1OUT    ; Selects the Pull-down resistor on P1.4

TEST:
bit.b      #00010000b, &P1IN     ; Testing pin P1.4
jnz        ON                    ; Jump if not zero: if a positive voltage is on pin
                                 ; P1.4 goto ON
jmp        TEST                  ; If still zero: goto TEST and keep on testing

ON:                              ; Whatever you want to happen when P1.4 has a
                                 ; positive voltage on it
```

Fig. 10.6 Test for logic high

test pin P1.4 for a situation where the pin is no longer grounded (i.e. a positive voltage has appeared on the pin). Reference Fig. 10.6.

10.3.3 Sample Code: Reading When the Switch Has Been Released (a "0") at P1.4

The following code snippet in Fig. 10.7 is added to sense when the switch is no longer pushed (no longer a logic high voltage on the pin). The first part of the code is the same as above.

```
mov.b      #00000000b, &P1SEL    ; Selects pins as GPIO and clears P1 Port
mov.b      #01000000b, &P1DIR    ; Selects pin P1.6 as OUTPUT, other pins are INPUT
mov.b      #10111111b, &P1REN    ; Enables Pull-up/Pull-down resistors on INPUT pins
mov.b      #10101111b, &P1OUT    ; Selects the Pull-down resistor on P1.4

TEST:
bit.b      #00010000b, &P1IN     ; Testing pin P1.4
jnz        GATE                  ; Jump if not zero …… if a positive voltage on pin
                                 ; P1.4 goto GATE
jmp        TEST                  ; If still zero …… goto TEST and keep on testing

GATE:
bit.b      #00010000b, &P1IN     ; Testing pin P1.4
jz         ON                    ; Jump if zero, if positive voltage removed from pin
                                 ; P1.4 goto ON
jmp        GATE                  ; If still not zero …… goto GATE and keep on testing

ON:                              ; Whatever you want to have happen when P1.4 has a
                                 ; positive voltage on it
```

Fig. 10.7 Test for logic low

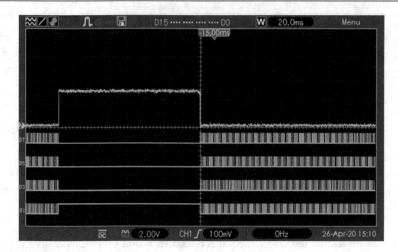

Fig. 10.8 Bounceless switch

Now there is only one pulse on the blue line when the tactile switch is pushed instead of many and continued scanning is paused until the switch is released. Reference Fig. 10.8.

10.3.4 Digital Lock

This project describes a novel approach for a security lock system. Figure 10.9 shows the MSP430G2553 microcontroller, the 4x3 Keypad, and an output board (1 Blue LED and 1 Red LED). The 4x3 Keypad is configured (and programmed) so that when a switch is pushed that number is entered into a register and saved. After five pushes of switches the Blue LED comes on signifying that there have been five pushes. If after the five pushes the numbers add up to 37 then the Red LED comes on signifying that the correct switches were pushed. It is possible to substitute a motor or solenoid for the Red LED.

As the program is written now there are several "5 push" combinations of numbers that add up to 37. To make the combination more secure, it is possible to make the magic number any number up to 65535 (maximum that a 16 bit register can hold) instead of 37. To accomplish this each switch would have to be assigned numbers other than 1–12. Reference Fig. 10.10.

Fig. 10.9 Digital lock

Example Switch Pushes: 3 + 5 + 8 + 10 + 11 = 37

1	2	3
4	5	6
7	8	9
10	11	12

a) Keypad Number assignment (as assembly program is written now).

Example Switch Pushes: 2107 + 32768 + 347 + 587 + 9235 = 45044

110	67	2107
598	32768	8923
75	347	8123
587	9235	28

b) Possible (more secure) keypad number assignment.

Fig. 10.10 Digital lock combinations

```
;-----------------------------------------------------------------------------
; Program Name: 4x3_SWITCH_MATRIX
; Assembly Code using Texas Instruments Code Composer
;
; Parts:
; -- Texas Instruments MSP-EXP430G2ET Launchpad with MSP430G2553
; -- 4x3 Switch Matrix (can be homemade)
; -- 2 LEDs for P2.1 and P2.5 (can be onboard Launchpad)

; Documentation:
; -- Texas Instruments SLAU144K Users Guide (ti.com)
; -- Texas Instruments SLAS735J MSP430G2553 Datasheet (ti.com)
;
; Connections:
;
;     MSP430G2553          4x3 Switch Matrix
;        P1.0                 Row #1
;        P1.1                 Row #2
;        P1.2                 Row #3
;        P1.3                 Row #4
;
```

```
;       P2.5                    Column #1
;       P2.6                    Column #2
;       P2.7                    Column #3
;
; This program is to be used with a 4x3 Keypad with 12 tactile switches. 4 rows
; across and 3 columns down. The clock is set for 1 MHz ... each row is brought
; "high" one at a time (P1.0, P1.1, P1.2, P1.3) and then each column is scanned
; for a "high" input on pins (P1.5, P1.6, P1.7) which will signify that a tactile
; switch was pushed. A pushed switch is not recognized until it is released. The
; program cycles through bringing the rows high and checking for inputs on the
; columns ... "scans" for pushes of the switches. In this particular program each
; switch is assigned a number (1-12) and when 5 switches have been pushed then the
; Blue LED (Pin P2.5) comes on signifying 5 switches have been pushed. If the 5
; pushes total 37 then the Red LED (Pin P2.1) comes on. If the total is not 37
; only the Blue LED comes on and the program starts over. This can be adapted for
; any numbers you want to assign to the switches. Also a solenoid, motor, or other
; device could be substituted for the Red LED.

; J. Kretzschmar
;-------------------------------------------------------------------------------
            .cdecls C,LIST,"msp430.h"       ; Include device header file
            .def    RESET
            .text
            .retain
            .retainrefs
;-------------------------------------------------------------------------------
RESET       mov.w   #__STACK_END,SP         ; Initialize stackpointer
StopWDT     mov.w   #WDTPW|WDTHOLD,&WDTCTL  ; Stop watchdog timer

        mov.b   &CALBC1_1MHZ, &BCSCTL1 ; Calibrated 1 MHz clock
        mov.b   &CALDCO_1MHZ, &DCOCTL  ; Calibrated 1 MHz clock
        mov.b   #48, &BCSCTL2          ; Clock divided by 8 (clock = 125 KHz)

        mov.b   #00000000b, &P1SEL ; Port_1 I/O port
        mov.b   #00001111b, &P1DIR ; Output bits: 0,1,2,3   Input bits: 4,5,6,7
        mov.b   #11110000b, &P1REN ; Pull-up/Pull-down resistors engaged on Inputs
        mov.b   #00000000b, &P1OUT ; Pull-down resistors engaged on bits 4,5,6,7
        mov.b   #00000000b, &P2SEL ; Port_2 I/O port
        mov.b   #11111111b, &P2DIR ; All bits are output bits
        bic.b   #11111111b, &P2OUT ; All bits are initially turned off
        mov     #0, R7
        mov     #0, R8


BEGIN_AGAIN:
; ------ SCANNING PROCESS ----------------------------------------------------
SCAN:
        bic.b   #00001111b, &P1OUT ; Output pins P1.0, P1.1, P1.2, P1.3 off
                                   ; Preserved Pull-down resistors P1.4 - P1.7
        bis.b   #00000001b, &P1OUT ; P1.0 brought HIGH
        bit.b   #00100000b, &P1IN  ; Input pin P1.5 tested if HIGH
        jnz     ONE_A             ; If HIGH goto ONE_A
        bit.b   #01000000b, &P1IN  ; Input pin P1.6 tested if HIGH
        jnz     TWO_A             ; If HIGH goto TWO_A
        bit.b   #10000000b, &P1IN  ; Input pin P1.7 tested if HIGH
        jnz     THREE_A           ; If HIGH goto THREE_A
        bic.b   #00000001b, &P1OUT ; P1.0 brought LOW

        bis.b   #00000010b, &P1OUT ; P1.1 brought HIGH
        bit.b   #00100000b, &P1IN  ; Input pin P1.5 tested if HIGH
```

```
        jnz     FOUR_A             ; If HIGH goto FOUR_A
        bit.b   #01000000b, &P1IN  ; Input pin P1.6 tested if HIGH
        jnz     FIVE_A             ; If HIGH goto FIVE_A
        bit.b   #10000000b, &P1IN  ; Input pin P1.7 tested if HIGH
        jnz     SIX_A              ; If HIGH goto SIX_A
        bic.b   #00000010b, &P1OUT ; P1.1 brought LOW

        bis.b   #00000100b, &P1OUT ; P1.2 brought HIGH
        bit.b   #00100000b, &P1IN  ; Input pin P1.5 tested if HIGH
        jnz     SEVEN_A            ; If HIGH goto SEVEN_A
        bit.b   #01000000b, &P1IN  ; Input pin P1.6 tested if HIGH
        jnz     EIGHT_A            ; If HIGH goto EIGHT_A
        bit.b   #10000000b, &P1IN  ; Input pin P1.7 tested if HIGH
        jnz     NINE_A             ; If HIGH goto NINE_A
        bic.b   #00000100b, &P1OUT ; P1.2 brought LOW

        bis.b   #00001000b, &P1OUT ; P1.3 brought HIGH
        bit.b   #00100000b, &P1IN  ; Input pin P1.5 tested if HIGH
        jnz     TEN_A              ; If HIGH goto TEN_A
        bit.b   #01000000b, &P1IN  ; Input pin P1.6 tested if HIGH
        jnz     ELEVEN_A           ; If HIGH goto ELEVEN_A
        bit.b   #10000000b, &P1IN  ; Input pin P1.7 tested if HIGH
        jnz     TWELVE_A           ; If HIGH goto TWELVE_A
        bic.b   #00001000b, &P1OUT ; P1.3 brought LOW


        jmp     SCAN               ; Goto SCAN and repeat process

; ------ THIS SECTION IS WHERE THE PROGRAM IS DIRECTED TO WHEN A SWITCH HAS --
; ------ BEEN PUSHED. NOW THE SWITCH IS TESTED TO SEE WHEN IT HAS BEEN -------
; ------ RELEASED. THIS ENSURES THAT ONLY ONE EVENT HAPPENS FOR ONE PUSH -----
; ------ OF THE SWITCH, OTHERWISE WITH THE SCANNING HAPPENING SO FAST --------
; ------ THE MICROCONTROLLER WOULD SENSE THAT MULTIPLE PUSHES OF THE SWITCH --
; ------ HAD OCCURRED AND THE PROGRAM WOULD NOT WORK AS INTENDED. ------------
; ------ SEE DOCUMENTATION FOR DETAILS --------------------------------------

ONE_A:
        bit.b   #00100000b, &P1IN  ; Input pin P1.5 tested if LOW
        jz      ONE                ; If LOW goto ONE
        jmp     ONE_A              ; If not LOW goto ONE_A and keep testing
ONE:
        add     #1, R7             ; Add 1 to R7 (R7 saves number up to 37)
        add     #1, R8             ; Add 1 to R8 (R8 counts for 5 pushes)
        cmp     #5, R8             ; Checks to see if 5 pushes have occurred
        jeq     PUSHES             ; If R8 = 5 then goto PUSHES
        jmp     SCAN               ; If R8 does not equal 5 continue scanning
TWO_A:
        bit.b   #01000000b, &P1IN  ; Input pin P1.6 tested if LOW
        jz      TWO
        jmp     TWO_A
TWO:
        add     #2, R7
        add     #1, R8
        cmp     #5, R8
        jeq     PUSHES
        jmp     SCAN
THREE_A:
        bit.b   #10000000b, &P1IN  ; Input pin P1.7 tested if LOW
        jz      THREE
        jmp     THREE_A
```

```
THREE:
       add     #3, R7
       add     #1, R8
       cmp     #5, R8
       jeq     PUSHES
       jmp     SCAN
FOUR_A:
       bit.b   #00100000b, &P1IN ; Input pin P1.5 tested if LOW
       jz      FOUR
       jmp     FOUR_A
FOUR:
       add     #4, R7
       add     #1, R8
       cmp     #5, R8
       jeq     PUSHES
       jmp     SCAN
FIVE_A:
       bit.b   #01000000b, &P1IN ; Input pin P1.6 tested if LOW
       jz      FIVE
       jmp     FIVE_A
FIVE:
       add     #5, R7
       add     #1, R8
       cmp     #5, R8
       jeq     PUSHES
       jmp     SCAN
SIX_A:
       bit.b   #10000000b, &P1IN ; Input pin P1.7 tested if LOW
       jz      SIX
       jmp     SIX_A
SIX:
       add     #6, R7
       add     #1, R8
       cmp     #5, R8
       jeq     PUSHES
       jmp     SCAN
SEVEN_A:
       bit.b   #00100000b, &P1IN ; Input pin P1.5 tested if LOW
       jz      SEVEN
       jmp     SEVEN_A
SEVEN:
       add     #7, R7
       add     #1, R8
       cmp     #5, R8
       jeq     PUSHES
       jmp     SCAN
EIGHT_A:
       bit.b   #01000000b, &P1IN ; Input pin P1.6 tested if LOW
       jz      EIGHT
       jmp     EIGHT_A
EIGHT:
       add     #8, R7
       add     #1, R8
       cmp     #5, R8
       jeq     PUSHES
       jmp     SCAN
NINE_A:
       bit.b   #10000000b, &P1IN ; Input pin P1.7 tested if LOW
       jz      NINE
```

```
        jmp       NINE_A
NINE:
        add       #9, R7
        add       #1, R8
        cmp       #5, R8
        jeq       PUSHES
        jmp       SCAN
TEN_A:
        bit.b     #00100000b, &P1IN ; Input pin P1.5 tested if LOW
        jz        TEN
        jmp       TEN_A
TEN:
        add       #10, R7
        add       #1, R8
        cmp       #5, R8
        jeq       PUSHES
        jmp       SCAN
ELEVEN_A:
        bit.b     #01000000b, &P1IN ; Input pin P1.6 tested if LOW
        jz        ELEVEN
        jmp       ELEVEN_A
ELEVEN:
        add       #11, R7
        add       #1, R8
        cmp       #5, R8
        jeq       PUSHES
        jmp       SCAN
TWELVE_A:
        bit.b     #10000000b, &P1IN ; Input pin P1.7 tested if LOW
        jz        TWELVE
        jmp       TWELVE_A
TWELVE:
        add       #12, R7
        add       #1, R8
        cmp       #5, R8
        jeq       PUSHES
        jmp       SCAN

; ------ THIS SECTION IS WHERE THE PROGRAM IS DIRECTED TO WHEN 5 PUSHES -----
; ------ OCCURRED. A SHORT DELAY IS SETUP UP SO THE BLUE LED STAYS ON -------
; ------ FOR A SHORT TIME SIGNIFYING THAT 5 PUSHES HAVE OCCURRED AND THE ----
; ------ PROGRAM NOW CHECKS TO SEE WHETHER OR NOT THE 5 PUSHES ADD UP TO ----
; ------ 37 ... IF THEY DO THEN RED LED COMES ON FOR A SHORT TIME (THIS -----
; ------ COULD ALSO BE A MOTOR OR SOLENOID THAT WOULD UNLOCK SOMETHING). ----
; ------ IF R8 DOES NOT EQUAL 37 THEN THE PROGRAM CLEARS R7 AND R8 AND ------
; ------ STARTS OVER AGAIN. -------------------------------------------------
PUSHES:
BLUE:
        bis.b     #00100000b, &P2OUT ; Turn on BLUE LED

        cmp       #37, R7           ; Does R7 equal 37
        jeq       UNLOCK            ; If so goto UNLOCK
                                    ; If R7 does not equal 37 continue on

        mov       #65535, R12       ; Delay to keep BLUE LED on
ON:
        dec       R12               ; Decrement R12
        jnz       ON                ; If not zero goto ON
```

```
        mov       #0, R7              ; Clear R7
        mov       #0, R8              ; Clear R8
        bic.b     #00100000b, &P2OUT  ; Turn off BLUE LED
        jmp       BEGIN_AGAIN         ; Start scanning again

UNLOCK:
        mov       #65535, R13         ; Delay
RED:
        bis.b     #00000010b, &P2OUT  ; Turn on RED LED
        dec       R13                 ; Decrement R13
        jnz       RED                 ; If R13 not zero goto RED
                                      ; If R13 equals zero continue on
        mov       #0, R7              ; Clear R7
        mov       #0, R8              ; Clear R8
        bic.b     #00100000b, &P2OUT  ; Turn off BLUE LED
        bic.b     #00000010b, &P2OUT  ; Turn off RED LED
        jmp       BEGIN_AGAIN         ; Start scanning again
;-----------------------------------------------------------------------
; Stack Pointer definition
;-----------------------------------------------------------------------
        .global  __STACK_END
        .sect    .stack
;-----------------------------------------------------------------------
; Interrupt Vectors
;-----------------------------------------------------------------------
        .sect    ".reset"                 ; MSP430 RESET Vector
        .short   RESET


;-----------------------------------------------------------------------
```

10.4   Project 3: Reading the Contents of a Register

This program demonstrates how the contents of a register can be read by using the Roll–Left–Through–Carry instruction, evaluating the Carry Bit, and blinking the GREEN LED on P1.0 if the Carry Bit holds a "0," and blinking the RED LED on P1.6 if the Carry Bit holds a "1." The clock was slowed down so that as the LEDs blink they can be written down starting with the most significant bit of the 16 bit register. Code Composer does have the capability of displaying the contents of a register, however, this demonstration shows another method of looking at the contents. Reference Fig. 10.11.

```
;-----------------------------------------------------------------------
; Program Name: READING_A_REGISTER
; Assembly Code using Texas Instruments Code Composer
;
; Parts:
; -- Texas Instruments MSP-EXP430G2ET Launchpad with MSP430G2553
;    with P1.0 and P1.6 LED jumper connectors in place
;
; Documentation:
; -- Texas Instruments SLAU144K Users Guide (ti.com)
; -- Texas Instruments SLAS735J MSP430G2553 Datasheet (ti.com)
;
```

Fig. 10.11 Reading contents of a register. Reading a register: GREEN LED (P1.0) = 0, Reading a register: RED LED (P1.6) = 1

```
; This program demonstrates how the contents of a register can be read by using
; the Roll-Left-Through-Carry instruction, evaluating the Carry Bit, and
; blinking the GREEN LED on P1.0 if the Carry Bit holds a "0", and blinking
; the RED LED on P1.6 if the Carry Bit holds a "1". The clock was slowed down
; so that as the LEDs blink they can be written down starting with the most
; significant bit of the 16 bit register. Code Composer does have the capability
; of displaying the contents of a register, however, this demonstration shows
; another method of looking at the contents.
;
; J. Kretzschmar
;
;-------------------------------------------------------------------------------
            .cdecls C,LIST,"msp430.h"       ; Include device header file
            .def    RESET
            .text
            .retain
            .retainrefs
;-------------------------------------------------------------------------------
RESET       mov.w   #__STACK_END,SP          ; Initialize stackpointer
StopWDT     mov.w   #WDTPW|WDTHOLD,&WDTCTL   ; Stop watchdog timer
;-------------------------------------------------------------------------------
;----- SLOW CLOCK SO BLINKS OF P1.0 AND P1.6 CAN BE WRITTEN DOWN ---------------
;-------------------------------------------------------------------------------
    mov     #32,&BCSCTL2                 ; Clock is Default at 1 MHz
```

```
                                      ; MCLK is DCO
                                      ; SMCLK is DCO
                                      ;  0 = MCLK/1 (Bits 4+5 of register "00")
                                      ; 16 = MCLK/2 (Bits 4+5 of register "01")
                                      ; 32 = MCLK/4 (Bits 4+5 of register "10")
                                      ; 48 = MCLK/8 (Bits 4+5 of register "11")

    mov.b    #01000001b, &P1DIR       ; P1.0 and P1.6 are output pins
    bic.b    #00000000b, &P1OUT       ; P1.0 and P1.6 are initially off
;---------------------------------------------------------------------------
;----- THIS IS FOR DEMONSTRATION PURPOSES ... NUMBER TO BE DISPLAYED ON --------
;----- P1.0 AND P1.6 IS LOADED INTO R6 (REGISTER 6) --------------------------
;---------------------------------------------------------------------------
    mov.w    #32657, R6               ; Number loaded into R6
;---------------------------------------------------------------------------
    mov.w    #16,R7                   ; Counter for the 16 times thru routine
READ_REGISTER:
    cmp      #0, R7                   ; Has counter hit zero yet?
    jeq      FINISH                   ; If counter at zero go to FINISH
    rlc      R6                       ; Roll one Bit left through Carry Bit
    jc       RED                      ; Carry Bit has a "1" ... goto RED
    jmp      GREEN                    ; Carry Bit has a "0" ... goto GREEN

RED:
    bis.b    #01000000b, &P1OUT       ; Turns RED LED on
    mov      #65535,R8
DELAY_0:                              ; Keeps RED LED on for a while
    dec      R8
    cmp      #0, R8
    jnz      DELAY_0
    bic.b    #01000000b, &P1OUT       ; Turns RED LED off
    mov      #65535,R8
DELAY_00:                             ; Keeps RED LED off for a while
    dec      R8
    jnz      DELAY_00
    dec      R7
    jmp      READ_REGISTER            ; Goto READ_REGISTER

GREEN:
    bis.b    #00000001b, &P1OUT       ; Turns GREEN LED on (P1.0)
    mov      #65535,R8
DELAY_1:                              ; Keeps GREEN LED on for a while
    dec      R8
    cmp      #0, R8
    jnz      DELAY_1
    bic.b    #00000001b, &P1OUT       ; Turns GREEN LED off (P1.0)
    mov      #65535,R8
DELAY_11:                             ; Keeps GREEN LED off for a while
    dec      R8
    jnz      DELAY_11
    dec      R7
    jmp      READ_REGISTER            ; Goto READ_REGISTER

FINISH:
    bic.b    #00000001b, &P1OUT       ; Turn off GREEN LED (P1.0)
    bic.b    #01000000b, &P1OUT       ; Turn off RED LED (P1.6)
    jmp      FINISH
;---------------------------------------------------------------------------
; Stack Pointer definition
```

```
;------------------------------------------------------------------------------
            .global  __STACK_END
            .sect    .stack
;------------------------------------------------------------------------------
; Interrupt Vectors
;------------------------------------------------------------------------------
            .sect    ".reset"                    ; MSP430 RESET Vector
            .short   RESET

;------------------------------------------------------------------------------
```

10.5   Project 4: Reading an LM34 Temperature Sensor

This project displays a temperature from an LM34 temperature sensor and demonstrates how a few assembly code instructions can be used to multiply and divide a number. The temperature sensor was tested around 70° and can display a temperature up to 99° (below 100 because of the math). The temperature is accurate to within a degree. Reference Fig. 10.12.

This program uses the ADC module on a MSP430G2553 microcontroller, inputs a voltage into channel 7 (P1.7) of the ADC from the LM34 temperature sensor, and processes the value which is then sent to the LCD display. A brief description about the math involved is provided in the comments of the code listing. LCD display delays are longer than noted in the datasheet.

```
;---------------------------------------------------------------
; Program Name: TEMPERATURE_SENSOR
; Assembly Code using Texas Instruments Code Composer
;
; Parts:
; -- Texas Instruments MSP-EXP430G2ET Launchpad with MSP430G2553
; -- Newhaven NHD-0216K1Z 16x2 LCD Display (or similar display)
; -- 10K Potentiometer (for contrast control)
; -- LM34 Temperature sensor
;
; Documentation:
; -- Texas Instruments SLAU144K Users Guide (ti.com)
; -- Texas Instruments SLAS735J MSP430G2553 Datasheet (ti.com)
; -- Newhaven NHD-0216K1Z Datasheet (newhavendisplay.com)
; -- LM34 Datasheet
; -- Paper "Controlling a 16x2 LCD Display with the MSP430G2553 Microcontroller
;    Using Assembly Code" (Appendix)
;
; This project displays a temperature from an LM34 temperature sensor and
; demonstrates how a few assembly code instructions can be used to multiply
; and divide a number. The temperature range was tested around 70 degrees and
; can reach 99 (below 100 because of the math). The temperature is accurate
; to within a degree.
;
; This program uses the ADC module on a MSP430G2553 microcontroller, inputs a
; voltage into channel 7 of the ADC (P1.7) from the LM34 temperature sensor
; which is then processed and sent to the LCD display. A brief description
; about the math: The LM34 outputs 10 millivolts per degree of temperature
```

Fig. 10.12 Reading an LM34 temperature sensor

```
; (example: 700 millivolts = 70 degrees). The ADC converts this voltage into
; a number somewhere between 0-1023. The reference voltage for the ADC is
; 3300 millivolts (3.3 volts), 1023 is the max number outputted from the ADC.
; Calculating the expected number from the ADC is accomplished by the math
; (1023 x 700 / 3300 = 217). The number 217 needs to be "processed" to
; represent the temperature. Using these factors: (1) 700 mV = 70 degrees
; is a factor of 10, (2) The ratio 700/217 = 3.22, (3) Multiplying a 16 bit
; binary number by 10 would require 5 shifts to the left. Every shift
; multiplies the number by 2, (4) 3.22 x 10 = 32.2 (disregard the 0.2). If
; the number outputted from the ADC is shifted 5 shifts to the left it is
; multiplied by 32. Accounting for the factor of 10 from the LM34 and the
; factor of 10 from 5 shifts, the actual temperature has been multiplied by
; 100. With this number, 100 can be subtracted from it with a counter
; that counts the number of times 100 has been subtracted from it. When the
; counter gets below 100 the temperature is then displayed on the LCD.
; LCD display delays are longer than noted in the datasheet.
;
; Connections:
;
;  LCD Display       MSP430G2553
;     Pin #1         Ground
;     Pin #2         +5 volts
;     Pin #3         Contrast control (center pin from 10K potentiometer)
;     Pin #4         P1.4 (R/S Register Select)
;     Pin #5         Ground (Read/Write selection: 0 for Write, 1 for Read)
;     Pin #6         P1.5 (Enable)
```

```
;     Pin #7            P2.0 (Data)
;     Pin #8            P2.1 (Data)
;     Pin #9            P2.2 (Data)
;     Pin #10           P2.3 (Data)
;     Pin #11           P2.4 (Data)
;     Pin #12           P2.5 (Data)
;     Pin #13           P2.6 (Data)
;     Pin #14           P2.7 (Data)
;     Pin #15           +5 volts (backlight)
;     Pin #16           Ground (backlight)
;
;                       P1.7 (output pin from LM34 temperature sensor)
;                       Ground (for LM34 temperature sensor)
;                       +5.0 volts (for LM34 temperature sensor)
;
; J. Kretzschmar


;-------------------------------------------------------------------------------
            .cdecls C,LIST,"msp430.h"      ; Include device header file
            .def    RESET
            .text
            .retain
            .retainrefs
;-------------------------------------------------------------------------------
RESET       mov.w   #__STACK_END,SP          ; Initialize stackpointer
StopWDT     mov.w   #WDTPW|WDTHOLD,&WDTCTL  ; Stop watchdog timer
;-------------------------------------------------------------------------------
;----- INITIAL SETUP MSP430G2553 PARAMETERS -----------------------------------
;-------------------------------------------------------------------------------
START_HERE:
    mov.b   &CALBC1_16MHZ,&BCSCTL1   ; Calibrated 16 MHz CPU Clock
    mov.b   &CALDCO_16MHZ,&DCOCTL    ; Calibrated 16 MHz CPU Clock

    mov.b   #00000000b, &P1SEL       ; Port 1 selected as I/O pins
    mov.b   #00000000b, &P2SEL       ; Port 2 selected as I/O pins
    mov.b   #00110000b, &P1DIR       ; P1.4 and P1.5 designated as OUTPUT pins
    mov.b   #11111111b, &P2DIR       ; P2.0 - P.7 designated as OUTPUT pins
    mov.b   #00000000b, &P1OUT       ; All Port_1 pins are cleared
    mov.b   #00000000b, &P2OUT       ; All Port_2 pins are cleared

    mov.w   #ADC10SHT_2+ADC10ON, &ADC10CTL0 ; Setting up ADC
    mov.w   #INCH_7, &ADC10CTL1             ; ADC input channel 7 (P1.7)

    bis.b   #BIT7, &ADC10AE0                ; Setting up ADC
;-------------------------------------------------------------------------------
;----- WAKE UP SEQUENCE OF LCD DISPLAY --- INITIAL 50 MILLISECOND DELAY ---------
;-------------------------------------------------------------------------------
    mov     #65535, R6          ; 12.5 Millisecond Delay
DELAY_0:
    dec     R6
    jnz     DELAY_0
    mov     #65535, R6          ; 12.5 Millisecond Delay
DELAY_00:
    dec     R6
    jnz     DELAY_00
    mov     #65535, R6          ; 12.5 Millisecond Delay
DELAY_000:
    dec     R6
    jnz     DELAY_000
```

```
     mov      #65535, R6           ; 12.5 Millisecond Delay
DELAY_0000:
     dec      R6
     jnz      DELAY_0000
;----- (COMMAND: FIRST WAKE UP CALL --- SELECTS 8-BIT, 2 LINE) -----------------
     mov.b #00111000b, R4       ; 8-Bit entry format, 2 Line
          call    #SETUP
;----- (COMMAND: SECOND WAKE UP CALL --- SELECTS 8-BIT, 2 LINE) ----------------
     mov.b    #00111000b, R4      ; 8-Bit entry format, 2 Line
     call     #SETUP
;----- (COMMAND: THIRD WAKE UP CALL --- SELECTS 8-BIT, 2 LINE) -----------------
     mov.b    #00111000b, R4      ; 8-Bit entry format, 2 Line
     call     #SETUP
;------------------------------------------------------------------------------
;----- SETUP INFORMATION TO BE ENTERED INTO THE LCD DISPLAY --------------------
;------------------------------------------------------------------------------
;----- (COMMAND: CLEAR DISPLAY) -----------------------------------------------
     mov.b    #00000001b, R4      ; Clear Display
     call     #SETUP
;----- (COMMAND: SETS DDRAM ADDRESS --- WHERE CHARACTERS APPEAR) ---------------
     mov.b    #10000111b, R4      ; DDRAM Address 07 selected
     call     #SETUP
;----- (COMMAND: DISPLAY CONTROL) ---------------------------------------------
     mov.b    #00001100b, R4      ; Display ON, Cursor OFF, Cursor Blinking OFF
     call     #SETUP
;----- (COMMAND: DISPLAY SHIFT/CURSOR MOVE (RIGHT OR LEFT) --------------------
     mov.b    #00000110b, R4      ; Display Shift OFF, Next character to the right
     call     #SETUP
;----- (COMMAND: CURSOR/DISPLAY SHIFT)(NOT IMPLEMENTED) ------------------------
     ;mov.b   #00010000b, R4     ; Cursor/Display Shift OFF, Shift Left
     ;call    #SETUP
;----- (COMMAND: CURSOR HOME)(NOT IMPLEMENTED) --------------------------------
     ;mov.b   #00000010b, R4      ; Cursor Home (DDRAM address 00)
     ;call    #SETUP
     jmp      INPUT_DATA_FOR_LABEL      ; All done with setup, goto Data Entry
;------------------------------------------------------------------------------
;----- SUBROUTINE THAT ENTERS SETUP INFORMATION INTO LCD DISPLAY ---------------
;------------------------------------------------------------------------------
SETUP:
     mov.b    #00000000b, &P1OUT  ; (P1.4 = 0) LCD DISPLAY Command Mode Selected

     bis.b    #00100000b, &P1OUT  ; (P1.5) ENABLE pin brought HIGH

     mov      #26215, R6           ; 5 Millisecond Delay
DELAY_1:
     dec      R6
     jnz      DELAY_1

     mov.b    R4, &P2OUT           ; Setup Information loaded into LCD DISPLAY

bic.b    #00100000b, &P1OUT  ; (P1.5) ENABLE pin brought LOW

     mov      #26215, R6           ; 5 Millisecond Delay
DELAY_2:
     dec      R6
     jnz      DELAY_2

     ret                          ; Returns program to after "call"
```

```
;-------------------------------------------------------------------------------
;----- THIS SECTION INPUTS DATA FOR LABEL AND ADC DATA -------------------------
;-------------------------------------------------------------------------------

INPUT_DATA_FOR_LABEL:

    mov.b   #10000010b, R4      ; Sets Location at DDRAM "02" for start of label
    call    #SETUP

    mov     #525, R12           ; 100 Microsecond Delay
DELAY_13:
    dec     R12
    jnz     DELAY_13

    mov     #84, R4             ; Loads ASCII number for "T"
    call    #SEND_ASCII_NUMBER_TO_LCD_DISPLAY

    mov     #101, R4            ; Loads ASCII number for "e"
    call    #SEND_ASCII_NUMBER_TO_LCD_DISPLAY

    mov     #109, R4            ; Loads ASCII number for "m"
    call    #SEND_ASCII_NUMBER_TO_LCD_DISPLAY

    mov     #112, R4            ; Loads ASCII number for "p"
    call    #SEND_ASCII_NUMBER_TO_LCD_DISPLAY

    mov     #101, R4            ; Loads ASCII number for "e"
    call    #SEND_ASCII_NUMBER_TO_LCD_DISPLAY

    mov     #114, R4            ; Loads ASCII number for "r"
    call    #SEND_ASCII_NUMBER_TO_LCD_DISPLAY

    mov     #97, R4             ; Loads ASCII number for "a"
    call    #SEND_ASCII_NUMBER_TO_LCD_DISPLAY

    mov     #116, R4            ; Loads ASCII number for "t"
    call    #SEND_ASCII_NUMBER_TO_LCD_DISPLAY

    mov     #117, R4            ; Loads ASCII number for "u"
    call    #SEND_ASCII_NUMBER_TO_LCD_DISPLAY

    mov     #114, R4            ; Loads ASCII number for "r"
    call    #SEND_ASCII_NUMBER_TO_LCD_DISPLAY

    mov     #101, R4            ; Loads ASCII number for "e"
    call    #SEND_ASCII_NUMBER_TO_LCD_DISPLAY

    mov.b   #11001000b, R4      ; Set Location at DDRAM "72"
    call    #SETUP              ; "Symbol for Degree"

    mov     #223, R4            ; Loads ASCII number for Degree Symbol
    call    #SEND_ASCII_NUMBER_TO_LCD_DISPLAY

    mov.b   #11001001b, R4      ; Set Location at DDRAM "73"
    call    #SETUP              ; "Symbol for Fahrenheit"

    mov     #70, R4             ; Loads ASCII number for "F"
    call    #SEND_ASCII_NUMBER_TO_LCD_DISPLAY
```

```
        jmp      PROCESS_ADC_DATA    ; Goto section to that processes ADC data

;--------------------------------------------------------------------------------
;----- PROCESSING OF ADC DATA SECTION -------------------------------------------
;--------------------------------------------------------------------------------

PROCESS_ADC_DATA:
;--------------------------------------------------------------------------------
;----- This section is a delay which consists of an Outer Delay Loop and an ------
;----- Inner Delay Loop. The Inner Delay Loop is approximately 4 Milliseconds ----
;----- and the Outer Delay Loop is how many times the Inner Delay Loop is run. ---
;----- The number can be changed as needed. With the CPU clock running at 16 MHz -
;----- and no delay the displayed numbers were "jumpy" and needed to be slowed ---
;----- down before they were sent to the LCD for display. -----------------------
;--------------------------------------------------------------------------------
LOOP:
        mov      #500, R14   ; This number can be changed as needed
OUTER_DELAY_LOOP:

        mov      #65535, R12 ; This number provides for a 4 Millisecond delay
INNER_DELAY_LOOP:
        dec      R12
        jnz      INNER_DELAY_LOOP

        dec      R14
        jnz      OUTER_DELAY_LOOP
;--------------------------------------------------------------------------------
;----------------------- END OF DELAY SECTION -----------------------------------
;--------------------------------------------------------------------------------
;
;--------------------------------------------------------------------------------
;----- THIS SECTION SETS DDRAM ADDRESS WHERE ADC DATA WILL BE DISPLAYED, ---------
;----- TURNS ON THE ADC TO ACQUIRE A BINARY NUMBER, CONVERTS THE BINARY NUMBER ---
;----- TO A BCD NUMBER, AND THEN CONVERTS THE BCD NUMBER TO AN ASCII NUMBER ------
;----- WHICH IS THEN SENT TO THE LCD TO BE DISPLAYED ----------------------------
;--------------------------------------------------------------------------------
        mov.b    #11000110b, R4 ; Sets location at DDRAM "70"
        call     #SETUP         ; This is where the temperature is displayed

        mov      #525, R12      ; 100 Microsecond Delay
DELAY_14:
        dec      R12
        jnz      DELAY_14
;--------------------------------------------------------------------------------
;----- TURNS ON ADC TO ACQUIRE A BINARY NUMBER ----------------------------------
;--------------------------------------------------------------------------------
        bis.w    #ENC+ADC10SC, &ADC10CTL0 ; Starts an ADC conversion

        mov.w    #60000, R8 ; Delay, gives time to settle before moving data
DELAY_15:
        dec      R8
        jnz      DELAY_15

        mov.w    &ADC10MEM, R5  ; Move data from ADC Memory Register to R5

;--------------------------------------------------------------------------------
;----- THIS SECTION TAKES THE BINARY NUMBER COMING OUT OF THE ADC AND MULTIPLIES ---
;----- BY 32 (5 SHIFTS TO THE LEFT). THIS WILL GIVE A DECIMAL NUMBER THAT WILL -----
;----- REPRESENT THE TEMPERATURE IN FAHRENHEIT, HOWEVER THIS NEW NUMBER NEEDS ------
```

```
;----- TO BE DIVIDED BY 100 FOR DISPLAY ON THE LCD. THE DIVISION IS ACCOMPLISHED ---
;----- BY A LOOP WHERE 100 IS SUBTRACTED FROM THE NUMBER AND THE NUMBER OF TIMES ---
;----- THROUGH THE LOOP IS ACCUMULATED IN A COUNTER THEN DISPLAYED ON THE LCD ------
;-------------------------------------------------------------------------------------
              rla.w    R5
              rla.w    R5
              rla.w    R5
              rla.w    R5
              rla.w    R5


SUBTRACTION_LOOP:
              add      #1, R9     ; R9 is the counter (Accumulator)
              cmp      #100, R5   ; Compare the number stored in R5 to 100
              jl       CONVERT_BINARY_NUMBER_TO_BCD_NUMBER ; If R5 is less than 100
                                  ; goto CONVERT_BINARY_NUMBER_TO_BCD_NUMBER
              sub      #100, R5   ; If R5 still greater than 100 then subtract 100
              jmp      SUBTRACTION_LOOP


;-------------------------------------------------------------------------------------
;----- SUBROUTINE TO CONVERT ADC BINARY NUMBER TO BCD NUMBER ----------------------
;-------------------------------------------------------------------------------------
CONVERT_BINARY_NUMBER_TO_BCD_NUMBER:
              mov      #0, R15    ; Make sure R15 is empty
              rla.w    R9         ; This section converts the binary number
              dadd.w   R15, R15   ; from the ADC into a BCD number. The
              rla.w    R9         ; assembly statement "dadd" that does this is
              dadd.w   R15, R15   ; only available for use when programming in
              rla.w    R9         ; assembly. These lines take the 16 Bit register
              dadd.w   R15, R15   ; and roll off one Bit at a time to the left
              rla.w    R9         ; and the "BCD Conversion" process is accomplished
              dadd.w   R15, R15   ; with "dadd" and stored in R15. These lines could
              rla.w    R9         ; be done with a loop process (x16), however it is
              dadd.w   R15, R15   ; important to see how the BCD conversion is done
              rla.w    R9
              dadd.w   R15, R15
              rla.w    R9
              dadd.w   R15, R15
              rla.w    R9
              dadd.w   R15, R15
              rla.w    R9
              dadd.w   R15, R15
              rla.w    R9
              dadd.w   R15, R15
              rla.w    R9
              dadd.w   R15, R15
              rla.w    R9
              dadd.w   R15, R15
              rla.w    R9
              dadd.w   R15, R15
              rla.w    R9
              dadd.w   R15, R15
              rla.w    R9
              dadd.w   R15, R15
;-------------------------------------------------------------------------------------
;----- READS BCD NUMBER IN BLOCKS OF 4 BITS ---------------------------------------
;-------------------------------------------------------------------------------------
              mov      #0, R7     ; Ensures R7 is empty
```

```
            mov       #2, R10   ; Place 2 in counter R10

            rla       R15       ; Since the temperature will only be dsiplaying
            rla       R15       ; temperture below 100 the upper 8 bits of R15
            rla       R15       ; only represent "0"s and do not need to be
            rla       R15       ; displayed on the LCD, therefore, these 8 bits
            rla       R15       ; are not evaluated.
            rla       R15
            rla       R15
            rla       R15

READ_4_BITS:                    ; This routine takes the BCD number stored in R15
            rlc       R15       ; and evaluates each block of 4 Bits one bit at a
            jc        EIGHT     ; time starting from the left. R7 stores the number
ROLL_BIT2:                      ; which is then converted to the appropriate ASCII
            rlc       R15       ; number and sent to be displayed on the LCD display
            jc        FOUR      ; This process is done 2 times so that the lower 8
ROLL_BIT1:                      ; bits of register (R15) are read (one BCD number
            rlc       R15       ; per 4 Bit block) ... then the process repeats with
            jc        TWO       ; a new number from the ADC
ROLL_BIT0:
            rlc       R15
            jc        ONE
            jmp       DIGIT

EIGHT:
            add       #8, R7
            jmp       ROLL_BIT2
FOUR:
            add       #4, R7
            jmp       ROLL_BIT1
TWO:
            add       #2, R7
            jmp       ROLL_BIT0
ONE:
            add       #1, R7
;-------------------------------------------------------------------------------
;----- TAKES BCD DIGIT AND CONVERTS IT TO ASCII NUMBER -------------------------
;-------------------------------------------------------------------------------
DIGIT:
            add       #48, R7   ; ASCII 48 = 0   Refer to ASCII table
            mov       R7, R4

            call      #SEND_ASCII_NUMBER_TO_LCD_DISPLAY

            mov       #0, R4    ; R4 cleared
            mov       #0, R7    ; R7 cleared
            dec       R10       ; 1 decremented from R10
            cmp       #0, R10   ; Does R10 have 0 stored in it
            jeq       LOOP      ; If yes, go get new ADC number
            jmp       READ_4_BITS ; If no, go get new 4-bit block
;-------------------------------------------------------------------------------
;----- SUBROUTINE THAT ENTERS ADC DATA (ASCII NUMBERS) INTO LCD DISPLAY ---------
;-------------------------------------------------------------------------------
SEND_ASCII_NUMBER_TO_LCD_DISPLAY:
    mov.b   #00010000b, &P1OUT  ; (P1.4 = 1) LCD DISPLAY Data Mode Selected

    bis.b   #00100000b, &P1OUT  ; (P1.5) ENABLE pin brought HIGH
```

```
        mov      #26215, R6          ; 5 Millisecond Delay
DELAY_7:
        dec      R6
        jnz      DELAY_7

        mov.b    R4, &P2OUT          ; ASCII Numbers (Data) loaded into LCD Display

bic.b    #00100000b, &P1OUT  ; P1.5 ENABLE pin brought LOW

        mov      #26215, R6          ; 5 Millisecond Delay
DELAY_8:
        dec      R6
        jnz      DELAY_8

        ret                          ; Returns program to after "call"
;------------------------------------------------------------------------------
; Stack Pointer definition
;------------------------------------------------------------------------------
            .global __STACK_END
            .sect   .stack
;------------------------------------------------------------------------------
; Interrupt Vectors
;------------------------------------------------------------------------------
            .sect   ".reset"                ; MSP430 RESET Vector
            .short  RESET
```

10.6 Post Lab Activities

1. In the Appendices we provide a brief introduction to Unified Modeling Language (UML) activity diagrams. These are UML compliant flow charts. Provide UML activity diagrams for programs provided in this laboratory exercise.
2. For assembly language programs provided; rewrite in C, compile, execute, and test for proper operation.

Appendix A
Introduction to Digilent Analog Discovery 2

A.1   Purpose

In this Appendix you will learn the basics on how to use the Digilent Analog Discovery 2 (DAD2) PC based test instrument. The DAD2, along with a host personal computer (PC) or laptop, provides a full suite of portable test equipment. The lab manual may also be used with traditional laboratory test equipment.

You will be using the oscilloscope function in several of the labs, and most likely the logic analyzer and waveform generator in future courses. The purpose of this lab is for you to become proficient in using this test instrument. Take the time to practice using the different controls and settings and explore this useful device.

A.2   Introduction

The Digilent Analog Discovery 2 is a PC–based multi–functional electronic test instrument that uses Digilent's Waveforms software. The card inside the box guides you through the software installation on your computer and the Digilent website provides several tutorials on the various functions of the Analog Discovery 2. In addition, the "Help" button will guide you through the initial learning of the instrument. Please see: www.digilent.com. The purpose of this lab exercise is to provide the necessary information to get started with the basic functions.

A.3   Overview

The Analog Discovery 2 "Bundle Package" comes with the Analog Discovery 2 pod, a USB cable, a BNC adapter board, two probes, and a 30–wire bundle (Flywire) as shown in

© The Editor(s) (if applicable) and The Author(s), under exclusive license to Springer Nature Switzerland AG 2023
J. Kretzschmar et al., *MSP430 Microcontroller Lab Manual*, Synthesis Lectures on Digital Circuits & Systems, https://doi.org/10.1007/978-3-031-26643-0

Fig. A.1. The BNC adapter board allows you to use probes for the Oscilloscope function and the Wave Generator function. The Flywire bundle allows you to have access to the functions without having to use the probes as well as provides inputs to the 16 channels of the Logic Analyzer. The card inside the box provides the information necessary to use the Flywire as shown in Fig. A.2.

Once the software has been loaded and the unit plugged into a USB port, the screen in Fig. A.3 will appear. The Analog Discovery 2 instrument has several functions, however, we will only concentrate on getting you started with the three functions circled in red in Fig. A.3, the Oscilloscope function ("Scope"), the Wave Generator ("Wavegen"), and the Logic Analyzer function ("Logic"). Once you have learned the basics then the more advanced functions can be investigated. We first setup the Wave Generator so it can be used as an input signal for the Oscilloscope.

Fig. A.1 Analog Discovery 2
bundle package

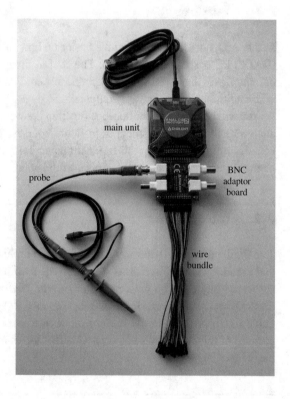

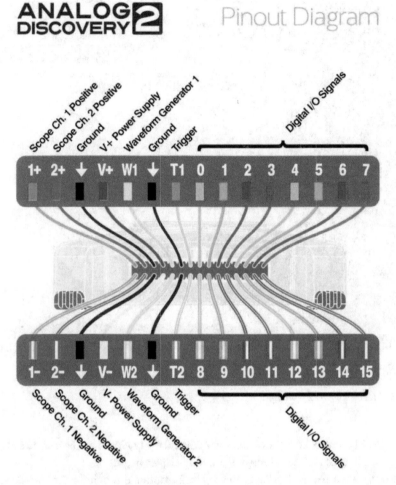

Fig. A.2 Analog Discovery 2 Flywire bundle (www.digilent.com)

A.4   Setting up the Wave Generator Function

Click on "Wavegen" and the window shown in Fig. A.4 appears. There are two Wave Generator channels. When this screen first comes up channel 1 is selected by default and channel 2 is off. For our basic test only channel 1 will be used, however, channel 2 can be selected if desired through the "Channels" button. The red circle shows where to select the type of wave, frequency, and amplitude. When you have made the selections you want, click on the "Run" button and the signal will be output on channel 1 of the Wave Generator. This signal is available from the BNC connector on the BNC Adapter board as well as the upper yellow

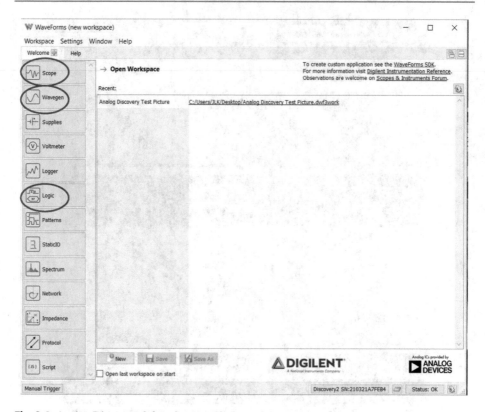

Fig. A.3 Analog Discovery 2 functions

wire on the Flywire bundle (however, only when the Flywire is connected directly to the Analog Discovery 2 pod, not through the BNC Adapter board).

The drop down menus for the frequency and amplitude provide a few pre-determined selections, however, any frequency or amplitude needed can be entered into the window and implemented by hitting the "Enter" button on your computer.

When using the Flywire for the output of the Wave Generator (or input for the Oscilloscope) it must be directly connected to the Analog Discovery 2 pod as shown if Fig. A.5, not through the BNC Adapter board. In the box is a small bag with five connecting pins. Attach one of the connectors as shown to the upper yellow wire of the Flywire (output for channel 1 of the Wave Generator), and to the black wire next to the yellow wire (ground connection for channel 1 of the Wave Generator). The Oscilloscope can be directly connected to the Wave Generator in this manner by connecting the upper orange wire (Oscilloscope channel 1) to the upper yellow wire (Wave Generator channel 1), and the 2 black wires (ground). Although both ground wires have the same internal connections in this particular situation, it is good practice to always make sure you have ground connections in place.

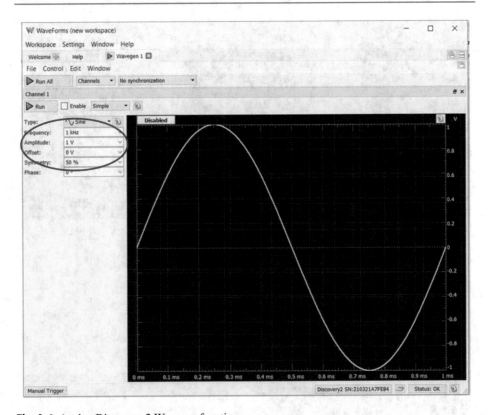

Fig. A.4 Analog Discovery 2 Wavegen function

A.5   Setting up the Oscilloscope Function

When using the BNC Adapter board to input a signal into the Oscilloscope two configurations have to be accomplished. On the BNC adapter board both oscilloscope channels need to have the jumper bar connected to "DC" and the oscilloscope probes need to be in the "1X" position as shown in Fig. A.6.

Next click on "Window" and then select "Waveforms" from the drop down menu. This will get you back to the main screen where you will now select "Scope" as shown in Fig. A.7.

Now both the Wave Generator and the Oscilloscope are open and you should see the screen shown in Fig. A.8 as depicted in the red circle.

Now that both the Wave Generator (channel 1) and the Oscilloscope (channel 1) are running, connect them together using one of the methods shown in Fig. A.9 (direct or through the BNC Adapter board).

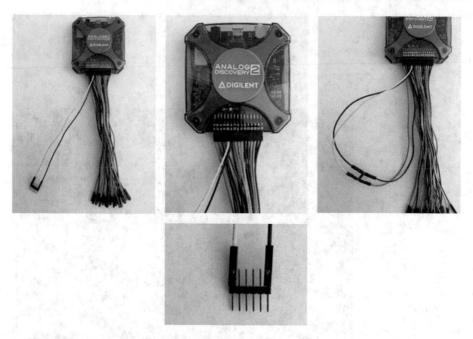

Fig. A.5 Analog Discovery 2 Flywire connections

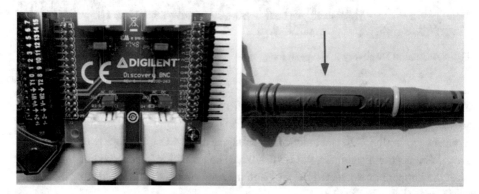

Fig. A.6 Analog Discovery 2 oscilloscope configuration

With the Wave Generator in the default settings of (Sine wave, 1.0 KHz, and 1.0 V amplitude) you will see the screen shown in Fig. A.10 when you open the Oscilloscope function. Initially both channel 1 and channel 2 will be operational and there will be a signal (yellow) on channel 1, and no signal (blue line) on channel 2. Changing the settings in the windows highlighted with red circles will change what is displayed on the scope.

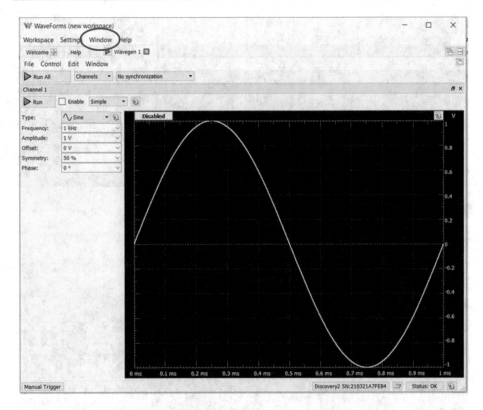

Fig. A.7 Analog Discovery 2 waveform generator

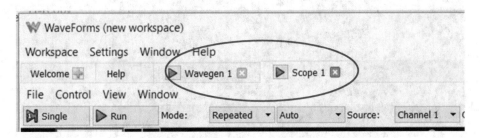

Fig. A.8 Analog Discovery 2 Wave Generator and Oscilloscope

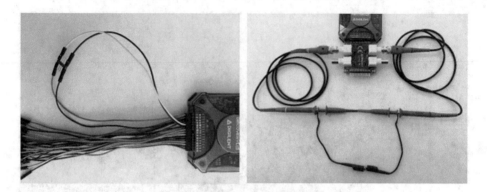

Fig. A.9 Analog Discovery 2 direct or via the BNC Adapter

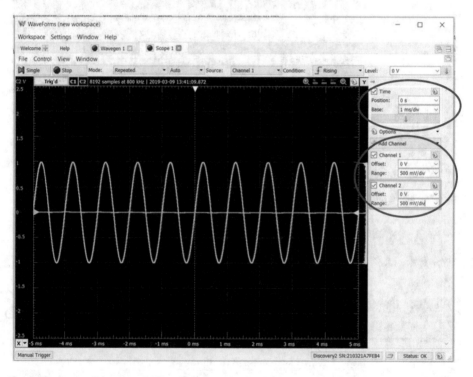

Fig. A.10 Analog Discovery 2 Wave Generator in the default settings of (Sine wave, 1.0 KHz, and 1.0 V amplitude)

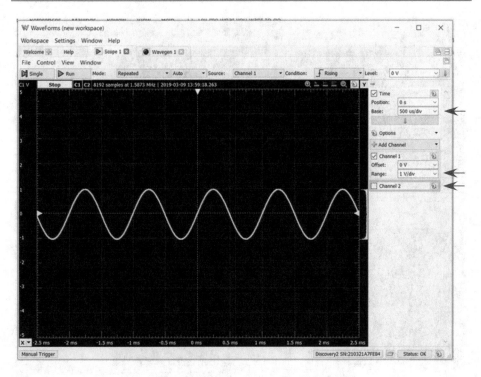

Fig. A.11 Analog Discovery 2 division adjustment

Horizontal time divisions can be changed by adjusting the Time Base in the drop down menu, and the vertical divisions can be changed by adjusting the Range in the drop down menu. Channel 2 can be turned off by unchecking the box and then the blue line will go away. Changes were made and are shown in Fig. A.11.

Measurement data can be obtained by going to the "View" button and selecting "Measurements" from the drop down menu. Clicking on "Measurements" opens up a new window, click on "Add" in the drop–down menu select "Defined Measurement." Next highlight the channel you want to capture measurements from, and then select "Vertical" and click on "Add," then select "Horizontal" and click on "Add." The measurements will be displayed. It may be necessary to expand the window on the left side to see all of the measurement data. The five pictures in Fig. A.12 show the process for obtaining measurement data.

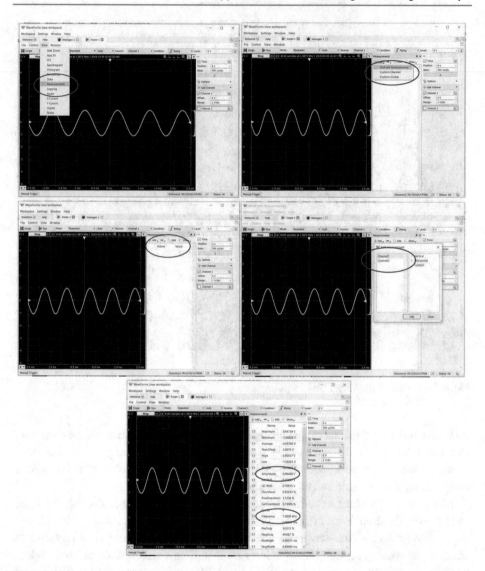

Fig. A.12 Measurement data process

A.6 Setting up the Logic Analyzer Function

When working with digital systems it is sometimes necessary to see the relationship between several signals at the same time. Whereas this instrument can have two input channels for the oscilloscope function (which many times is sufficient), the logic analyzer has 16 input channels. We show you how to accomplish the basic setup for the logic analyzer. From the main menu click on "Logic" as shown in Fig. A.13.

This next window allows you to add the channels that you want to view. Click on "Click to Add channels" and select "Signal," then you will be able to select the channels for inputting signals into the logic analyzer as shown in Fig. A.14. When selecting the channels hold down the "control" button on your computer.

Channels 3, 4, 5, 6, and 7 have been chosen to display five output signals (Fig. A.15). The five output signals can be generated by test code that turns the pins 3, 4, 5, 6, and 7 on and off in particular patterns. There is room to display 11 more channels if needed, and the horizontal time division markings can be changed using the drop–down menu highlighted with the red circle.

Fig. A.13 Analog Discovery 2 logic analyzer function

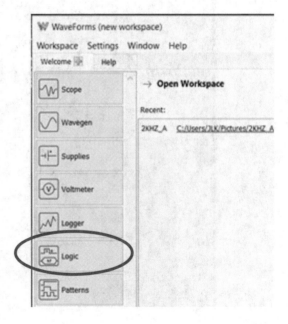

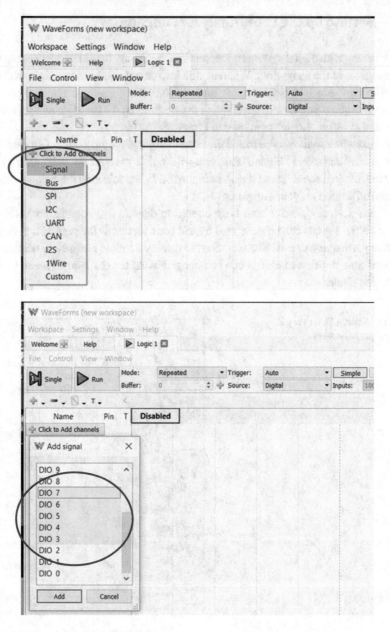

Fig. A.14 Logic analyzer channel selection

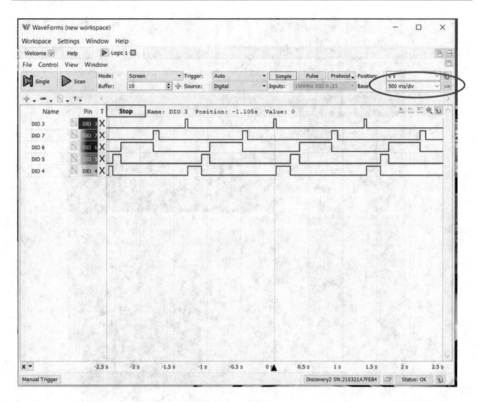

Fig. A.15 Logic analyzer channel selection. The five output signals are generated by test code that turns the pins 3, 4, 5, 6, and 7 on and off in particular patterns

---

A.7   Saving Images of the Screen

Occasionally you may want to save an image of the screen to incorporate into a report; the following is the process you will use to accomplish this task. The steps are illustrated in Fig. A.16.

1. Stop the function you are running and click on "File." A drop–down menu will appear and select "Export."
2. Click on "Save."
3. Select where you want your screenshot saved and give the file a name. Your file will be saved as a ".jpeg" file.

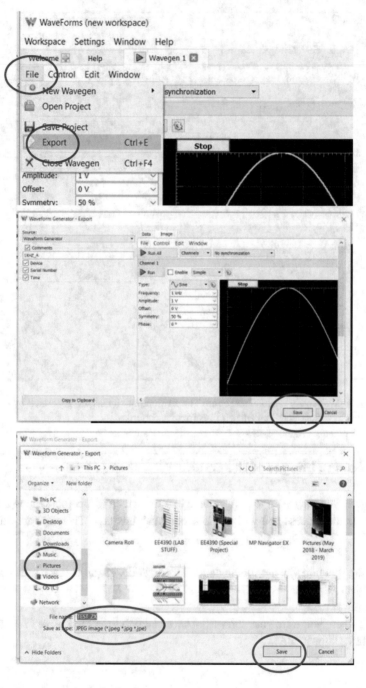

Fig. A.16 Logic analyzer channel selection

Appendix B
Distance Learning Lab Kit

This appendix illustrates the equipment and parts that are used to complete the laboratory exercises described in this book (Figs. B.1, B.2, B.3, B.4, B.5, B.6, B.7, B.8, B.9 and B.10).

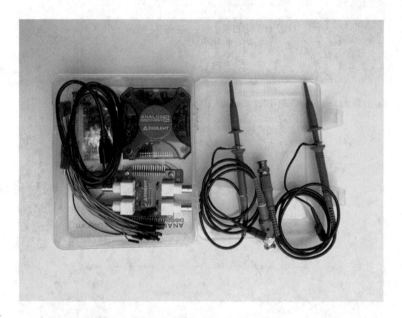

Fig. B.1 Digilent Analog Discovery 2 Test Instrument

© The Editor(s) (if applicable) and The Author(s), under exclusive license to Springer
Nature Switzerland AG 2023
J. Kretzschmar et al., *MSP430 Microcontroller Lab Manual*, Synthesis Lectures on Digital
Circuits & Systems, https://doi.org/10.1007/978-3-031-26643-0

Fig. B.2 Texas Instruments MSP–EXP430G2ET LaunchPads

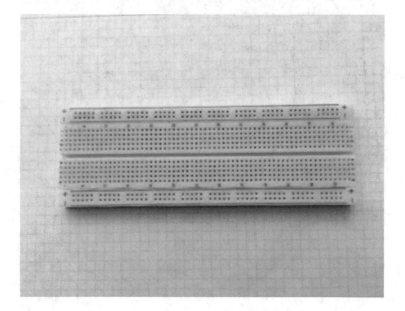

Fig. B.3 Solderless Breadboard

Fig. B.4 Jumper Wires (3 types: M–M, M–F, F–F)

Fig. B.5 LCD Display

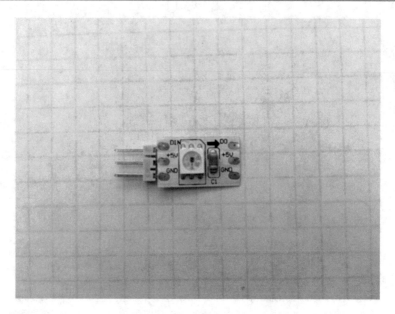

Fig. B.6 RGB (Red–Green–Blue) LED (WS2812B). The LED is very bright. Recommend taping a piece of paper over the LED

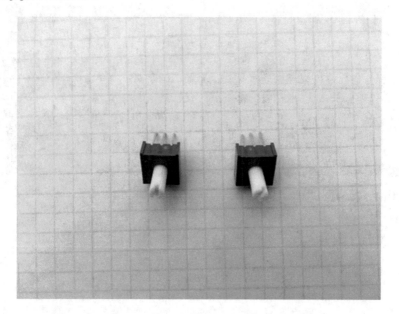

Fig. B.7 10K Potentiometers

Fig. B.8 Tactile Pushbutton Switches

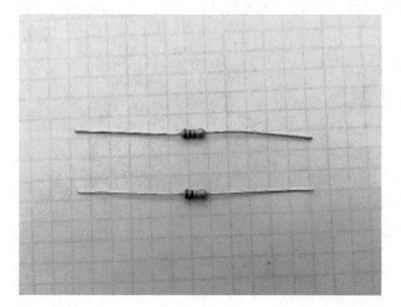

Fig. B.9 10K Resistors

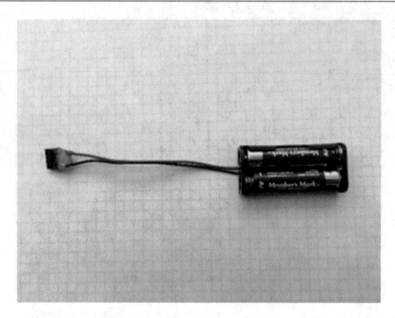

Fig. B.10 Battery Holder (AAA x2)

Appendix C
MSP430G2553 Header File

```
/* ============================================================================= */
/* Copyright (c) 2015, Texas Instruments Incorporated                            */
/*  All rights reserved.                                                         */
/*                                                                               */
/*  Redistribution and use in source and binary forms, with or without          */
/*  modification, are permitted provided that the following conditions           */
/*  are met:                                                                     */
/*                                                                               */
/*  *  Redistributions of source code must retain the above copyright            */
/*     notice, this list of conditions and the following disclaimer.             */
/*                                                                               */
/*  *  Redistributions in binary form must reproduce the above copyright         */
/*     notice, this list of conditions and the following disclaimer in the       */
/*     documentation and/or other materials provided with the distribution.      */
/*                                                                               */
/*  *  Neither the name of Texas Instruments Incorporated nor the names of       */
/*     its contributors may be used to endorse or promote products derived       */
/*     from this software without specific prior written permission.             */
/*                                                                               */
/*  THIS SOFTWARE IS PROVIDED BY THE COPYRIGHT HOLDERS AND CONTRIBUTORS "AS IS"  */
/*  AND ANY EXPRESS OR IMPLIED WARRANTIES, INCLUDING, BUT NOT LIMITED TO,        */
/*  THE IMPLIED WARRANTIES OF MERCHANTABILITY AND FITNESS FOR A PARTICULAR       */
/*  PURPOSE ARE DISCLAIMED. IN NO EVENT SHALL THE COPYRIGHT OWNER OR             */
/*  CONTRIBUTORS BE LIABLE FOR ANY DIRECT, INDIRECT, INCIDENTAL, SPECIAL,        */
/*  EXEMPLARY, OR CONSEQUENTIAL DAMAGES (INCLUDING, BUT NOT LIMITED TO,          */
/*  PROCUREMENT OF SUBSTITUTE GOODS OR SERVICES; LOSS OF USE, DATA, OR PROFITS;  */
/*  OR BUSINESS INTERRUPTION) HOWEVER CAUSED AND ON ANY THEORY OF LIABILITY,     */
/*  WHETHER IN CONTRACT, STRICT LIABILITY, OR TORT (INCLUDING NEGLIGENCE OR      */
/*  OTHERWISE) ARISING IN ANY WAY OUT OF THE USE OF THIS SOFTWARE,               */
/*  EVEN IF ADVISED OF THE POSSIBILITY OF SUCH DAMAGE.                           */
/* ============================================================================= */

/********************************************************************
*
* Standard register and bit definitions for the Texas Instruments
* MSP430 microcontroller.
*
* This file supports assembler and C development for
* MSP430G2553 devices.
*
* Texas Instruments, Version 1.2
*
* Rev. 1.0, Setup
* Rev. 1.1, added additional Cal Data Labels
* Rev. 1.2, added dummy TRAPINT_VECTOR interrupt vector as bugfix for USCI29
*
********************************************************************/
```

© The Editor(s) (if applicable) and The Author(s), under exclusive license to Springer   141
Nature Switzerland AG 2023
J. Kretzschmar et al., *MSP430 Microcontroller Lab Manual*, Synthesis Lectures on Digital
Circuits & Systems, https://doi.org/10.1007/978-3-031-26643-0

```
#ifndef __MSP430G2553
#define __MSP430G2553

#define __MSP430_HEADER_VERSION__ 1159

#define __MSP430_TI_HEADERS__

#ifdef __cplusplus
extern "C" {
#endif

#include <iomacros.h>

/************************************************************
* STANDARD BITS
************************************************************/

#define BIT0                (0x0001)
#define BIT1                (0x0002)
#define BIT2                (0x0004)
#define BIT3                (0x0008)
#define BIT4                (0x0010)
#define BIT5                (0x0020)
#define BIT6                (0x0040)
#define BIT7                (0x0080)
#define BIT8                (0x0100)
#define BIT9                (0x0200)
#define BITA                (0x0400)
#define BITB                (0x0800)
#define BITC                (0x1000)
#define BITD                (0x2000)
#define BITE                (0x4000)
#define BITF                (0x8000)

/************************************************************
* STATUS REGISTER BITS
************************************************************/

#define C                   (0x0001)
#define Z                   (0x0002)
#define N                   (0x0004)
#define V                   (0x0100)
#define GIE                 (0x0008)
#define CPUOFF              (0x0010)
#define OSCOFF              (0x0020)
#define SCG0                (0x0040)
#define SCG1                (0x0080)

/* Low Power Modes coded with Bits 4-7 in SR */

#ifndef __STDC__ /* Begin #defines for assembler */
#define LPM0                (CPUOFF)
#define LPM1                (SCG0+CPUOFF)
#define LPM2                (SCG1+CPUOFF)
#define LPM3                (SCG1+SCG0+CPUOFF)
#define LPM4                (SCG1+SCG0+OSCOFF+CPUOFF)
/* End #defines for assembler */

#else /* Begin #defines for C */
#define LPM0_bits           (CPUOFF)
#define LPM1_bits           (SCG0+CPUOFF)
#define LPM2_bits           (SCG1+CPUOFF)
#define LPM3_bits           (SCG1+SCG0+CPUOFF)
#define LPM4_bits           (SCG1+SCG0+OSCOFF+CPUOFF)

#include "in430.h"

#define LPM0      __bis_SR_register(LPM0_bits)         /* Enter Low Power Mode 0 */
#define LPM0_EXIT __bic_SR_register_on_exit(LPM0_bits) /* Exit Low Power Mode 0 */
```

```
#define LPM1        __bis_SR_register(LPM1_bits)         /* Enter Low Power Mode 1 */
#define LPM1_EXIT __bic_SR_register_on_exit(LPM1_bits) /* Exit Low Power Mode 1 */
#define LPM2        __bis_SR_register(LPM2_bits)         /* Enter Low Power Mode 2 */
#define LPM2_EXIT __bic_SR_register_on_exit(LPM2_bits) /* Exit Low Power Mode 2 */
#define LPM3        __bis_SR_register(LPM3_bits)         /* Enter Low Power Mode 3 */
#define LPM3_EXIT __bic_SR_register_on_exit(LPM3_bits) /* Exit Low Power Mode 3 */
#define LPM4        __bis_SR_register(LPM4_bits)         /* Enter Low Power Mode 4 */
#define LPM4_EXIT __bic_SR_register_on_exit(LPM4_bits) /* Exit Low Power Mode 4 */
#endif /* End #defines for C */

/************************************************************
* PERIPHERAL FILE MAP
************************************************************/

/************************************************************
* SPECIAL FUNCTION REGISTER ADDRESSES + CONTROL BITS
************************************************************/

#define IE1_                0x0000    /* Interrupt Enable 1 */
sfrb(IE1, IE1_);
#define WDTIE               (0x01)    /* Watchdog Interrupt Enable */
#define OFIE                (0x02)    /* Osc. Fault  Interrupt Enable */
#define NMIIE               (0x10)    /* NMI Interrupt Enable */
#define ACCVIE              (0x20)    /* Flash Access Violation Interrupt Enable */

#define IFG1_               0x0002    /* Interrupt Flag 1 */
sfrb(IFG1, IFG1_);
#define WDTIFG              (0x01)    /* Watchdog Interrupt Flag */
#define OFIFG               (0x02)    /* Osc. Fault Interrupt Flag */
#define PORIFG              (0x04)    /* Power On Interrupt Flag */
#define RSTIFG              (0x08)    /* Reset Interrupt Flag */
#define NMIIFG              (0x10)    /* NMI Interrupt Flag */

#define IE2_                0x0001    /* Interrupt Enable 2 */
sfrb(IE2, IE2_);
#define UC0IE               IE2
#define UCA0RXIE            (0x01)
#define UCA0TXIE            (0x02)
#define UCB0RXIE            (0x04)
#define UCB0TXIE            (0x08)

#define IFG2_               0x0003    /* Interrupt Flag 2 */
sfrb(IFG2, IFG2_);
#define UC0IFG              IFG2
#define UCA0RXIFG           (0x01)
#define UCA0TXIFG           (0x02)
#define UCB0RXIFG           (0x04)
#define UCB0TXIFG           (0x08)


/************************************************************
* ADC10
************************************************************/
#define __MSP430_HAS_ADC10__          /* Definition to show that Module is available */

#define ADC10DTC0_          0x0048    /* ADC10 Data Transfer Control 0 */
sfrb(ADC10DTC0, ADC10DTC0_);
#define ADC10DTC1_          0x0049    /* ADC10 Data Transfer Control 1 */
sfrb(ADC10DTC1, ADC10DTC1_);
#define ADC10AE0_           0x004A    /* ADC10 Analog Enable 0 */
sfrb(ADC10AE0, ADC10AE0_);

#define ADC10CTL0_          0x01B0    /* ADC10 Control 0 */
sfrw(ADC10CTL0, ADC10CTL0_);
#define ADC10CTL1_          0x01B2    /* ADC10 Control 1 */
sfrw(ADC10CTL1, ADC10CTL1_);
#define ADC10MEM_           0x01B4    /* ADC10 Memory */
sfrw(ADC10MEM, ADC10MEM_);
#define ADC10SA_            0x01BC    /* ADC10 Data Transfer Start Address */
```

```
sfrw(ADC10SA, ADC10SA_);

/* ADC10CTL0 */
#define ADC10SC             (0x001)     /* ADC10 Start Conversion */
#define ENC                 (0x002)     /* ADC10 Enable Conversion */
#define ADC10IFG            (0x004)     /* ADC10 Interrupt Flag */
#define ADC10IE             (0x008)     /* ADC10 Interrupt Enalbe */
#define ADC10ON             (0x010)     /* ADC10 On/Enable */
#define REFON               (0x020)     /* ADC10 Reference on */
#define REF2_5V             (0x040)     /* ADC10 Ref 0:1.5V / 1:2.5V */
#define MSC                 (0x080)     /* ADC10 Multiple SampleConversion */
#define REFBURST            (0x100)     /* ADC10 Reference Burst Mode */
#define REFOUT              (0x200)     /* ADC10 Enalbe output of Ref. */
#define ADC10SR             (0x400)     /* ADC10 Sampling Rate 0:200ksps / 1:50ksps */
#define ADC10SHT0           (0x800)     /* ADC10 Sample Hold Select Bit: 0 */
#define ADC10SHT1           (0x1000)    /* ADC10 Sample Hold Select Bit: 1 */
#define SREF0               (0x2000)    /* ADC10 Reference Select Bit: 0 */
#define SREF1               (0x4000)    /* ADC10 Reference Select Bit: 1 */
#define SREF2               (0x8000)    /* ADC10 Reference Select Bit: 2 */
#define ADC10SHT_0          (0x0000)    /* 4 x ADC10CLKs */
#define ADC10SHT_1          (0x0800)    /* 8 x ADC10CLKs */
#define ADC10SHT_2          (0x1000)    /* 16 x ADC10CLKs */
#define ADC10SHT_3          (0x1800)    /* 64 x ADC10CLKs */

#define SREF_0              (0x0000)    /* VR+ = AVCC and VR- = AVSS */
#define SREF_1              (0x2000)    /* VR+ = VREF+ and VR- = AVSS */
#define SREF_2              (0x4000)    /* VR+ = VEREF+ and VR- = AVSS */
#define SREF_3              (0x6000)    /* VR+ = VEREF+ and VR- = AVSS */
#define SREF_4              (0x8000)    /* VR+ = AVCC and VR- = VREF-/VEREF- */
#define SREF_5              (0xA000)    /* VR+ = VREF+ and VR- = VREF-/VEREF- */
#define SREF_6              (0xC000)    /* VR+ = VEREF+ and VR- = VREF-/VEREF- */
#define SREF_7              (0xE000)    /* VR+ = VEREF+ and VR- = VREF-/VEREF- */

/* ADC10CTL1 */
#define ADC10BUSY           (0x0001)    /* ADC10 BUSY */
#define CONSEQ0             (0x0002)    /* ADC10 Conversion Sequence Select 0 */
#define CONSEQ1             (0x0004)    /* ADC10 Conversion Sequence Select 1 */
#define ADC10SSEL0          (0x0008)    /* ADC10 Clock Source Select Bit: 0 */
#define ADC10SSEL1          (0x0010)    /* ADC10 Clock Source Select Bit: 1 */
#define ADC10DIV0           (0x0020)    /* ADC10 Clock Divider Select Bit: 0 */
#define ADC10DIV1           (0x0040)    /* ADC10 Clock Divider Select Bit: 1 */
#define ADC10DIV2           (0x0080)    /* ADC10 Clock Divider Select Bit: 2 */
#define ISSH                (0x0100)    /* ADC10 Invert Sample Hold Signal */
#define ADC10DF             (0x0200)    /* ADC10 Data Format 0:binary 1:2's complement */
#define SHS0                (0x0400)    /* ADC10 Sample/Hold Source Bit: 0 */
#define SHS1                (0x0800)    /* ADC10 Sample/Hold Source Bit: 1 */
#define INCH0               (0x1000)    /* ADC10 Input Channel Select Bit: 0 */
#define INCH1               (0x2000)    /* ADC10 Input Channel Select Bit: 1 */
#define INCH2               (0x4000)    /* ADC10 Input Channel Select Bit: 2 */
#define INCH3               (0x8000)    /* ADC10 Input Channel Select Bit: 3 */

#define CONSEQ_0            (0x0000)       /* Single channel single conversion */
#define CONSEQ_1            (0x0002)       /* Sequence of channels */
#define CONSEQ_2            (0x0004)       /* Repeat single channel */
#define CONSEQ_3            (0x0006)       /* Repeat sequence of channels */

#define ADC10SSEL_0         (0x0000)       /* ADC10OSC */
#define ADC10SSEL_1         (0x0008)       /* ACLK */
#define ADC10SSEL_2         (0x0010)       /* MCLK */
#define ADC10SSEL_3         (0x0018)       /* SMCLK */

#define ADC10DIV_0          (0x0000)    /* ADC10 Clock Divider Select 0 */
#define ADC10DIV_1          (0x0020)    /* ADC10 Clock Divider Select 1 */
#define ADC10DIV_2          (0x0040)    /* ADC10 Clock Divider Select 2 */
#define ADC10DIV_3          (0x0060)    /* ADC10 Clock Divider Select 3 */
#define ADC10DIV_4          (0x0080)    /* ADC10 Clock Divider Select 4 */
#define ADC10DIV_5          (0x00A0)    /* ADC10 Clock Divider Select 5 */
#define ADC10DIV_6          (0x00C0)    /* ADC10 Clock Divider Select 6 */
#define ADC10DIV_7          (0x00E0)    /* ADC10 Clock Divider Select 7 */
```

```
#define SHS_0                (0x0000)    /* ADC10SC */
#define SHS_1                (0x0400)    /* TA3 OUT1 */
#define SHS_2                (0x0800)    /* TA3 OUT0 */
#define SHS_3                (0x0C00)    /* TA3 OUT2 */

#define INCH_0               (0x0000)  /* Selects Channel 0 */
#define INCH_1               (0x1000)  /* Selects Channel 1 */
#define INCH_2               (0x2000)  /* Selects Channel 2 */
#define INCH_3               (0x3000)  /* Selects Channel 3 */
#define INCH_4               (0x4000)  /* Selects Channel 4 */
#define INCH_5               (0x5000)  /* Selects Channel 5 */
#define INCH_6               (0x6000)  /* Selects Channel 6 */
#define INCH_7               (0x7000)  /* Selects Channel 7 */
#define INCH_8               (0x8000)  /* Selects Channel 8 */
#define INCH_9               (0x9000)  /* Selects Channel 9 */
#define INCH_10              (0xA000) /* Selects Channel 10 */
#define INCH_11              (0xB000) /* Selects Channel 11 */
#define INCH_12              (0xC000) /* Selects Channel 12 */
#define INCH_13              (0xD000) /* Selects Channel 13 */
#define INCH_14              (0xE000) /* Selects Channel 14 */
#define INCH_15              (0xF000) /* Selects Channel 15 */

/* ADC10DTC0 */
#define ADC10FETCH           (0x001)    /* This bit should normally be reset */
#define ADC10B1              (0x002)    /* ADC10 block one */
#define ADC10CT              (0x004)    /* ADC10 continuous transfer */
#define ADC10TB              (0x008)    /* ADC10 two-block mode */
#define ADC10DISABLE         (0x000)    /* ADC10DTC1 */

/************************************************************
* Basic Clock Module
************************************************************/
#define __MSP430_HAS_BC2__            /* Definition to show that Module is available */

#define DCOCTL_              0x0056    /* DCO Clock Frequency Control */
sfrb(DCOCTL, DCOCTL_);
#define BCSCTL1_             0x0057    /* Basic Clock System Control 1 */
sfrb(BCSCTL1, BCSCTL1_);
#define BCSCTL2_             0x0058    /* Basic Clock System Control 2 */
sfrb(BCSCTL2, BCSCTL2_);
#define BCSCTL3_             0x0053    /* Basic Clock System Control 3 */
sfrb(BCSCTL3, BCSCTL3_);

#define MOD0                 (0x01)    /* Modulation Bit 0 */
#define MOD1                 (0x02)    /* Modulation Bit 1 */
#define MOD2                 (0x04)    /* Modulation Bit 2 */
#define MOD3                 (0x08)    /* Modulation Bit 3 */
#define MOD4                 (0x10)    /* Modulation Bit 4 */
#define DCO0                 (0x20)    /* DCO Select Bit 0 */
#define DCO1                 (0x40)    /* DCO Select Bit 1 */
#define DCO2                 (0x80)    /* DCO Select Bit 2 */

#define RSEL0                (0x01)    /* Range Select Bit 0 */
#define RSEL1                (0x02)    /* Range Select Bit 1 */
#define RSEL2                (0x04)    /* Range Select Bit 2 */
#define RSEL3                (0x08)    /* Range Select Bit 3 */
#define DIVA0                (0x10)    /* ACLK Divider 0 */
#define DIVA1                (0x20)    /* ACLK Divider 1 */
#define XTS                  (0x40)    /* LFXTCLK 0:Low Freq. / 1: High Freq. */
#define XT2OFF               (0x80)    /* Enable XT2CLK */

#define DIVA_0               (0x00)    /* ACLK Divider 0: /1 */
#define DIVA_1               (0x10)    /* ACLK Divider 1: /2 */
#define DIVA_2               (0x20)    /* ACLK Divider 2: /4 */
#define DIVA_3               (0x30)    /* ACLK Divider 3: /8 */

#define DIVS0                (0x02)    /* SMCLK Divider 0 */
#define DIVS1                (0x04)    /* SMCLK Divider 1 */
```

```
#define SELS               (0x08)   /* SMCLK Source Select 0:DCOCLK / 1:XT2CLK/LFXTCLK */
#define DIVM0              (0x10)   /* MCLK Divider 0 */
#define DIVM1              (0x20)   /* MCLK Divider 1 */
#define SELM0              (0x40)   /* MCLK Source Select 0 */
#define SELM1              (0x80)   /* MCLK Source Select 1 */

#define DIVS_0             (0x00)   /* SMCLK Divider 0: /1 */
#define DIVS_1             (0x02)   /* SMCLK Divider 1: /2 */
#define DIVS_2             (0x04)   /* SMCLK Divider 2: /4 */
#define DIVS_3             (0x06)   /* SMCLK Divider 3: /8 */

#define DIVM_0             (0x00)   /* MCLK Divider 0: /1 */
#define DIVM_1             (0x10)   /* MCLK Divider 1: /2 */
#define DIVM_2             (0x20)   /* MCLK Divider 2: /4 */
#define DIVM_3             (0x30)   /* MCLK Divider 3: /8 */

#define SELM_0             (0x00)   /* MCLK Source Select 0: DCOCLK */
#define SELM_1             (0x40)   /* MCLK Source Select 1: DCOCLK */
#define SELM_2             (0x80)   /* MCLK Source Select 2: XT2CLK/LFXTCLK */
#define SELM_3             (0xC0)   /* MCLK Source Select 3: LFXTCLK */

#define LFXT1OF            (0x01)   /* Low/high Frequency Oscillator Fault Flag */
#define XT2OF              (0x02)   /* High frequency oscillator 2 fault flag */
#define XCAP0              (0x04)   /* XIN/XOUT Cap 0 */
#define XCAP1              (0x08)   /* XIN/XOUT Cap 1 */
#define LFXT1S0            (0x10)   /* Mode 0 for LFXT1 (XTS = 0) */
#define LFXT1S1            (0x20)   /* Mode 1 for LFXT1 (XTS = 0) */
#define XT2S0              (0x40)   /* Mode 0 for XT2 */
#define XT2S1              (0x80)   /* Mode 1 for XT2 */

#define XCAP_0             (0x00)   /* XIN/XOUT Cap : 0 pF */
#define XCAP_1             (0x04)   /* XIN/XOUT Cap : 6 pF */
#define XCAP_2             (0x08)   /* XIN/XOUT Cap : 10 pF */
#define XCAP_3             (0x0C)   /* XIN/XOUT Cap : 12.5 pF */

#define LFXT1S_0           (0x00)   /* Mode 0 for LFXT1 : Normal operation */
#define LFXT1S_1           (0x10)   /* Mode 1 for LFXT1 : Reserved */
#define LFXT1S_2           (0x20)   /* Mode 2 for LFXT1 : VLO */
#define LFXT1S_3           (0x30)   /* Mode 3 for LFXT1 : Digital input signal */

#define XT2S_0             (0x00)   /* Mode 0 for XT2 : 0.4 - 1 MHz */
#define XT2S_1             (0x40)   /* Mode 1 for XT2 : 1 - 4 MHz */
#define XT2S_2             (0x80)   /* Mode 2 for XT2 : 2 - 16 MHz */
#define XT2S_3             (0xC0)   /* Mode 3 for XT2 : Digital input signal */

/************************************************************
* Comparator A
************************************************************/
#define __MSP430_HAS_CAPLUS__        /* Definition to show that Module is available */

#define CACTL1_            0x0059   /* Comparator A Control 1 */
sfrb(CACTL1, CACTL1_);
#define CACTL2_            0x005A   /* Comparator A Control 2 */
sfrb(CACTL2, CACTL2_);
#define CAPD_              0x005B   /* Comparator A Port Disable */
sfrb(CAPD, CAPD_);

#define CAIFG              (0x01)   /* Comp. A Interrupt Flag */
#define CAIE               (0x02)   /* Comp. A Interrupt Enable */
#define CAIES              (0x04)   /* Comp. A Int. Edge Select: 0:rising / 1:falling */
#define CAON               (0x08)   /* Comp. A enable */
#define CAREF0             (0x10)   /* Comp. A Internal Reference Select 0 */
#define CAREF1             (0x20)   /* Comp. A Internal Reference Select 1 */
#define CARSEL             (0x40)   /* Comp. A Internal Reference Enable */
#define CAEX               (0x80)   /* Comp. A Exchange Inputs */

#define CAREF_0            (0x00)   /* Comp. A Int. Ref. Select 0 : Off */
#define CAREF_1            (0x10)   /* Comp. A Int. Ref. Select 1 : 0.25*Vcc */
#define CAREF_2            (0x20)   /* Comp. A Int. Ref. Select 2 : 0.5*Vcc */
```

```
#define CAREF_3              (0x30)    /* Comp. A Int. Ref. Select 3 : Vt*/

#define CAOUT                (0x01)    /* Comp. A Output */
#define CAF                  (0x02)    /* Comp. A Enable Output Filter */
#define P2CA0                (0x04)    /* Comp. A +Terminal Multiplexer */
#define P2CA1                (0x08)    /* Comp. A -Terminal Multiplexer */
#define P2CA2                (0x10)    /* Comp. A -Terminal Multiplexer */
#define P2CA3                (0x20)    /* Comp. A -Terminal Multiplexer */
#define P2CA4                (0x40)    /* Comp. A +Terminal Multiplexer */
#define CASHORT              (0x80)    /* Comp. A Short + and - Terminals */

#define CAPD0                (0x01)    /* Comp. A Disable Input Buffer of Port Register .0 */
#define CAPD1                (0x02)    /* Comp. A Disable Input Buffer of Port Register .1 */
#define CAPD2                (0x04)    /* Comp. A Disable Input Buffer of Port Register .2 */
#define CAPD3                (0x08)    /* Comp. A Disable Input Buffer of Port Register .3 */
#define CAPD4                (0x10)    /* Comp. A Disable Input Buffer of Port Register .4 */
#define CAPD5                (0x20)    /* Comp. A Disable Input Buffer of Port Register .5 */
#define CAPD6                (0x40)    /* Comp. A Disable Input Buffer of Port Register .6 */
#define CAPD7                (0x80)    /* Comp. A Disable Input Buffer of Port Register .7 */

/************************************************************
* Flash Memory
************************************************************/
#define __MSP430_HAS_FLASH2__        /* Definition to show that Module is available */

#define FCTL1_               0x0128   /* FLASH Control 1 */
sfrw(FCTL1, FCTL1_);
#define FCTL2_               0x012A   /* FLASH Control 2 */
sfrw(FCTL2, FCTL2_);
#define FCTL3_               0x012C   /* FLASH Control 3 */
sfrw(FCTL3, FCTL3_);

#define FRKEY                (0x9600) /* Flash key returned by read */
#define FWKEY                (0xA500) /* Flash key for write */
#define FXKEY                (0x3300) /* for use with XOR instruction */

#define ERASE                (0x0002) /* Enable bit for Flash segment erase */
#define MERAS                (0x0004) /* Enable bit for Flash mass erase */
#define WRT                  (0x0040) /* Enable bit for Flash write */
#define BLKWRT               (0x0080) /* Enable bit for Flash segment write */
#define SEGWRT               (0x0080) /* old definition */ /* Enable bit for Flash segment write */

#define FN0                  (0x0001) /* Divide Flash clock by 1 to 64 using FN0 to FN5 according to: */
#define FN1                  (0x0002) /*  32*FN5 + 16*FN4 + 8*FN3 + 4*FN2 + 2*FN1 + FN0 + 1 */
#ifndef FN2
#define FN2                  (0x0004)
#endif
#ifndef FN3
#define FN3                  (0x0008)
#endif
#ifndef FN4
#define FN4                  (0x0010)
#endif
#define FN5                  (0x0020)
#define FSSEL0               (0x0040) /* Flash clock select 0 */   /* to distinguish from USART SSELx */
#define FSSEL1               (0x0080) /* Flash clock select 1 */

#define FSSEL_0              (0x0000) /* Flash clock select: 0 - ACLK */
#define FSSEL_1              (0x0040) /* Flash clock select: 1 - MCLK */
#define FSSEL_2              (0x0080) /* Flash clock select: 2 - SMCLK */
#define FSSEL_3              (0x00C0) /* Flash clock select: 3 - SMCLK */

#define BUSY                 (0x0001) /* Flash busy: 1 */
#define KEYV                 (0x0002) /* Flash Key violation flag */
#define ACCVIFG              (0x0004) /* Flash Access violation flag */
#define WAIT                 (0x0008) /* Wait flag for segment write */
#define LOCK                 (0x0010) /* Lock bit: 1 - Flash is locked (read only) */
#define EMEX                 (0x0020) /* Flash Emergency Exit */
#define LOCKA                (0x0040) /* Segment A Lock bit: read = 1 - Segment is locked (read only) */
```

```
#define FAIL                    (0x0080)  /* Last Program or Erase failed */

/************************************************************
* DIGITAL I/O Port1/2 Pull up / Pull down Resistors
************************************************************/
#define __MSP430_HAS_PORT1_R__          /* Definition to show that Module is available */
#define __MSP430_HAS_PORT2_R__          /* Definition to show that Module is available */

#define P1IN_                   0x0020    /* Port 1 Input */
const_sfrb(P1IN, P1IN_);
#define P1OUT_                  0x0021    /* Port 1 Output */
sfrb(P1OUT, P1OUT_);
#define P1DIR_                  0x0022    /* Port 1 Direction */
sfrb(P1DIR, P1DIR_);
#define P1IFG_                  0x0023    /* Port 1 Interrupt Flag */
sfrb(P1IFG, P1IFG_);
#define P1IES_                  0x0024    /* Port 1 Interrupt Edge Select */
sfrb(P1IES, P1IES_);
#define P1IE_                   0x0025    /* Port 1 Interrupt Enable */
sfrb(P1IE, P1IE_);
#define P1SEL_                  0x0026    /* Port 1 Selection */
sfrb(P1SEL, P1SEL_);
#define P1SEL2_                 0x0041    /* Port 1 Selection 2 */
sfrb(P1SEL2, P1SEL2_);
#define P1REN_                  0x0027    /* Port 1 Resistor Enable */
sfrb(P1REN, P1REN_);

#define P2IN_                   0x0028    /* Port 2 Input */
const_sfrb(P2IN, P2IN_);
#define P2OUT_                  0x0029    /* Port 2 Output */
sfrb(P2OUT, P2OUT_);
#define P2DIR_                  0x002A    /* Port 2 Direction */
sfrb(P2DIR, P2DIR_);
#define P2IFG_                  0x002B    /* Port 2 Interrupt Flag */
sfrb(P2IFG, P2IFG_);
#define P2IES_                  0x002C    /* Port 2 Interrupt Edge Select */
sfrb(P2IES, P2IES_);
#define P2IE_                   0x002D    /* Port 2 Interrupt Enable */
sfrb(P2IE, P2IE_);
#define P2SEL_                  0x002E    /* Port 2 Selection */
sfrb(P2SEL, P2SEL_);
#define P2SEL2_                 0x0042    /* Port 2 Selection 2 */
sfrb(P2SEL2, P2SEL2_);
#define P2REN_                  0x002F    /* Port 2 Resistor Enable */
sfrb(P2REN, P2REN_);

/************************************************************
* DIGITAL I/O Port3 Pull up / Pull down Resistors
************************************************************/
#define __MSP430_HAS_PORT3_R__          /* Definition to show that Module is available */

#define P3IN_                   0x0018    /* Port 3 Input */
const_sfrb(P3IN, P3IN_);
#define P3OUT_                  0x0019    /* Port 3 Output */
sfrb(P3OUT, P3OUT_);
#define P3DIR_                  0x001A    /* Port 3 Direction */
sfrb(P3DIR, P3DIR_);
#define P3SEL_                  0x001B    /* Port 3 Selection */
sfrb(P3SEL, P3SEL_);
#define P3SEL2_                 0x0043    /* Port 3 Selection 2 */
sfrb(P3SEL2, P3SEL2_);
#define P3REN_                  0x0010    /* Port 3 Resistor Enable */
sfrb(P3REN, P3REN_);

/************************************************************
* Timer0_A3
************************************************************/
#define __MSP430_HAS_TA3__              /* Definition to show that Module is available */
```

```
#define TA0IV_               0x012E    /* Timer0_A3 Interrupt Vector Word */
const_sfrw(TA0IV, TA0IV_);
#define TA0CTL_              0x0160    /* Timer0_A3 Control */
sfrw(TA0CTL, TA0CTL_);
#define TA0CCTL0_            0x0162    /* Timer0_A3 Capture/Compare Control 0 */
sfrw(TA0CCTL0, TA0CCTL0_);
#define TA0CCTL1_            0x0164    /* Timer0_A3 Capture/Compare Control 1 */
sfrw(TA0CCTL1, TA0CCTL1_);
#define TA0CCTL2_            0x0166    /* Timer0_A3 Capture/Compare Control 2 */
sfrw(TA0CCTL2, TA0CCTL2_);
#define TA0R_                0x0170    /* Timer0_A3 Counter Register */
sfrw(TA0R, TA0R_);
#define TA0CCR0_             0x0172    /* Timer0_A3 Capture/Compare 0 */
sfrw(TA0CCR0, TA0CCR0_);
#define TA0CCR1_             0x0174    /* Timer0_A3 Capture/Compare 1 */
sfrw(TA0CCR1, TA0CCR1_);
#define TA0CCR2_             0x0176    /* Timer0_A3 Capture/Compare 2 */
sfrw(TA0CCR2, TA0CCR2_);

/* Alternate register names */
#define TAIV                TA0IV     /* Timer A Interrupt Vector Word */
#define TACTL               TA0CTL    /* Timer A Control */
#define TACCTL0             TA0CCTL0  /* Timer A Capture/Compare Control 0 */
#define TACCTL1             TA0CCTL1  /* Timer A Capture/Compare Control 1 */
#define TACCTL2             TA0CCTL2  /* Timer A Capture/Compare Control 2 */
#define TAR                 TA0R      /* Timer A Counter Register */
#define TACCR0              TA0CCR0   /* Timer A Capture/Compare 0 */
#define TACCR1              TA0CCR1   /* Timer A Capture/Compare 1 */
#define TACCR2              TA0CCR2   /* Timer A Capture/Compare 2 */
#define TAIV_               TA0IV_    /* Timer A Interrupt Vector Word */
#define TACTL_              TA0CTL_   /* Timer A Control */
#define TACCTL0_            TA0CCTL0_ /* Timer A Capture/Compare Control 0 */
#define TACCTL1_            TA0CCTL1_ /* Timer A Capture/Compare Control 1 */
#define TACCTL2_            TA0CCTL2_ /* Timer A Capture/Compare Control 2 */
#define TAR_                TA0R_     /* Timer A Counter Register */
#define TACCR0_             TA0CCR0_  /* Timer A Capture/Compare 0 */
#define TACCR1_             TA0CCR1_  /* Timer A Capture/Compare 1 */
#define TACCR2_             TA0CCR2_  /* Timer A Capture/Compare 2 */

/* Alternate register names 2 */
#define CCTL0               TACCTL0   /* Timer A Capture/Compare Control 0 */
#define CCTL1               TACCTL1   /* Timer A Capture/Compare Control 1 */
#define CCTL2               TACCTL2   /* Timer A Capture/Compare Control 2 */
#define CCR0                TACCR0    /* Timer A Capture/Compare 0 */
#define CCR1                TACCR1    /* Timer A Capture/Compare 1 */
#define CCR2                TACCR2    /* Timer A Capture/Compare 2 */
#define CCTL0_              TACCTL0_  /* Timer A Capture/Compare Control 0 */
#define CCTL1_              TACCTL1_  /* Timer A Capture/Compare Control 1 */
#define CCTL2_              TACCTL2_  /* Timer A Capture/Compare Control 2 */
#define CCR0_               TACCR0_   /* Timer A Capture/Compare 0 */
#define CCR1_               TACCR1_   /* Timer A Capture/Compare 1 */
#define CCR2_               TACCR2_   /* Timer A Capture/Compare 2 */

#define TASSEL1             (0x0200)  /* Timer A clock source select 1 */
#define TASSEL0             (0x0100)  /* Timer A clock source select 0 */
#define ID1                 (0x0080)  /* Timer A clock input divider 1 */
#define ID0                 (0x0040)  /* Timer A clock input divider 0 */
#define MC1                 (0x0020)  /* Timer A mode control 1 */
#define MC0                 (0x0010)  /* Timer A mode control 0 */
#define TACLR               (0x0004)  /* Timer A counter clear */
#define TAIE                (0x0002)  /* Timer A counter interrupt enable */
#define TAIFG               (0x0001)  /* Timer A counter interrupt flag */

#define MC_0                (0x0000)  /* Timer A mode control: 0 - Stop */
#define MC_1                (0x0010)  /* Timer A mode control: 1 - Up to CCR0 */
#define MC_2                (0x0020)  /* Timer A mode control: 2 - Continous up */
#define MC_3                (0x0030)  /* Timer A mode control: 3 - Up/Down */
#define ID_0                (0x0000)  /* Timer A input divider: 0 - /1 */
#define ID_1                (0x0040)  /* Timer A input divider: 1 - /2 */
```

```
#define ID_2                (0x0080)  /* Timer A input divider: 2 - /4 */
#define ID_3                (0x00C0)  /* Timer A input divider: 3 - /8 */
#define TASSEL_0            (0x0000)  /* Timer A clock source select: 0 - TACLK */
#define TASSEL_1            (0x0100)  /* Timer A clock source select: 1 - ACLK  */
#define TASSEL_2            (0x0200)  /* Timer A clock source select: 2 - SMCLK */
#define TASSEL_3            (0x0300)  /* Timer A clock source select: 3 - INCLK */

#define CM1                 (0x8000)  /* Capture mode 1 */
#define CM0                 (0x4000)  /* Capture mode 0 */
#define CCIS1               (0x2000)  /* Capture input select 1 */
#define CCIS0               (0x1000)  /* Capture input select 0 */
#define SCS                 (0x0800)  /* Capture sychronize */
#define SCCI                (0x0400)  /* Latched capture signal (read) */
#define CAP                 (0x0100)  /* Capture mode: 1 /Compare mode : 0 */
#define OUTMOD2             (0x0080)  /* Output mode 2 */
#define OUTMOD1             (0x0040)  /* Output mode 1 */
#define OUTMOD0             (0x0020)  /* Output mode 0 */
#define CCIE                (0x0010)  /* Capture/compare interrupt enable */
#define CCI                 (0x0008)  /* Capture input signal (read) */
#define OUT                 (0x0004)  /* PWM Output signal if output mode 0 */
#define COV                 (0x0002)  /* Capture/compare overflow flag */
#define CCIFG               (0x0001)  /* Capture/compare interrupt flag */

#define OUTMOD_0            (0x0000)  /* PWM output mode: 0 - output only */
#define OUTMOD_1            (0x0020)  /* PWM output mode: 1 - set */
#define OUTMOD_2            (0x0040)  /* PWM output mode: 2 - PWM toggle/reset */
#define OUTMOD_3            (0x0060)  /* PWM output mode: 3 - PWM set/reset */
#define OUTMOD_4            (0x0080)  /* PWM output mode: 4 - toggle */
#define OUTMOD_5            (0x00A0)  /* PWM output mode: 5 - Reset */
#define OUTMOD_6            (0x00C0)  /* PWM output mode: 6 - PWM toggle/set */
#define OUTMOD_7            (0x00E0)  /* PWM output mode: 7 - PWM reset/set */
#define CCIS_0              (0x0000)  /* Capture input select: 0 - CCIxA */
#define CCIS_1              (0x1000)  /* Capture input select: 1 - CCIxB */
#define CCIS_2              (0x2000)  /* Capture input select: 2 - GND */
#define CCIS_3              (0x3000)  /* Capture input select: 3 - Vcc */
#define CM_0                (0x0000)  /* Capture mode: 0 - disabled */
#define CM_1                (0x4000)  /* Capture mode: 1 - pos. edge */
#define CM_2                (0x8000)  /* Capture mode: 1 - neg. edge */
#define CM_3                (0xC000)  /* Capture mode: 1 - both edges */

/* T0_A3IV Definitions */
#define TA0IV_NONE          (0x0000)    /* No Interrupt pending */
#define TA0IV_TACCR1        (0x0002)    /* TA0CCR1_CCIFG */
#define TA0IV_TACCR2        (0x0004)    /* TA0CCR2_CCIFG */
#define TA0IV_6             (0x0006)    /* Reserved */
#define TA0IV_8             (0x0008)    /* Reserved */
#define TA0IV_TAIFG         (0x000A)    /* TA0IFG */

/************************************************************
* Timer1_A3
************************************************************/
#define __MSP430_HAS_T1A3__         /* Definition to show that Module is available */

#define TA1IV_             0x011E   /* Timer1_A3 Interrupt Vector Word */
const_sfrw(TA1IV, TA1IV_);
#define TA1CTL_            0x0180   /* Timer1_A3 Control */
sfrw(TA1CTL, TA1CTL_);
#define TA1CCTL0_          0x0182   /* Timer1_A3 Capture/Compare Control 0 */
sfrw(TA1CCTL0, TA1CCTL0_);
#define TA1CCTL1_          0x0184   /* Timer1_A3 Capture/Compare Control 1 */
sfrw(TA1CCTL1, TA1CCTL1_);
#define TA1CCTL2_          0x0186   /* Timer1_A3 Capture/Compare Control 2 */
sfrw(TA1CCTL2, TA1CCTL2_);
#define TA1R_              0x0190   /* Timer1_A3 Counter Register */
sfrw(TA1R, TA1R_);
#define TA1CCR0_           0x0192   /* Timer1_A3 Capture/Compare 0 */
sfrw(TA1CCR0, TA1CCR0_);
#define TA1CCR1_           0x0194   /* Timer1_A3 Capture/Compare 1 */
sfrw(TA1CCR1, TA1CCR1_);
```

```
#define TA1CCR2_              0x0196    /* Timer1_A3 Capture/Compare 2 */
sfrw(TA1CCR2, TA1CCR2_);

/* Bits are already defined within the Timer0_Ax */

/* T1_A3IV Definitions */
#define TA1IV_NONE            (0x0000)  /* No Interrupt pending */
#define TA1IV_TACCR1          (0x0002)  /* TA1CCR1_CCIFG */
#define TA1IV_TACCR2          (0x0004)  /* TA1CCR2_CCIFG */
#define TA1IV_6               (0x0006)  /* Reserved */
#define TA1IV_8               (0x0008)  /* Reserved */
#define TA1IV_TAIFG           (0x000A)  /* TA1IFG */

/************************************************************
* USCI
************************************************************/
#define __MSP430_HAS_USCI__             /* Definition to show that Module is available */

#define UCA0CTL0_             0x0060    /* USCI A0 Control Register 0 */
sfrb(UCA0CTL0, UCA0CTL0_);
#define UCA0CTL1_             0x0061    /* USCI A0 Control Register 1 */
sfrb(UCA0CTL1, UCA0CTL1_);
#define UCA0BR0_              0x0062    /* USCI A0 Baud Rate 0 */
sfrb(UCA0BR0, UCA0BR0_);
#define UCA0BR1_              0x0063    /* USCI A0 Baud Rate 1 */
sfrb(UCA0BR1, UCA0BR1_);
#define UCA0MCTL_            0x0064    /* USCI A0 Modulation Control */
sfrb(UCA0MCTL, UCA0MCTL_);
#define UCA0STAT_            0x0065    /* USCI A0 Status Register */
sfrb(UCA0STAT, UCA0STAT_);
#define UCA0RXBUF_           0x0066    /* USCI A0 Receive Buffer */
const_sfrb(UCA0RXBUF, UCA0RXBUF_);
#define UCA0TXBUF_           0x0067    /* USCI A0 Transmit Buffer */
sfrb(UCA0TXBUF, UCA0TXBUF_);
#define UCA0ABCTL_           0x005D    /* USCI A0 LIN Control */
sfrb(UCA0ABCTL, UCA0ABCTL_);
#define UCA0IRTCTL_          0x005E    /* USCI A0 IrDA Transmit Control */
sfrb(UCA0IRTCTL, UCA0IRTCTL_);
#define UCA0IRRCTL_          0x005F    /* USCI A0 IrDA Receive Control */
sfrb(UCA0IRRCTL, UCA0IRRCTL_);


#define UCB0CTL0_            0x0068    /* USCI B0 Control Register 0 */
sfrb(UCB0CTL0, UCB0CTL0_);
#define UCB0CTL1_            0x0069    /* USCI B0 Control Register 1 */
sfrb(UCB0CTL1, UCB0CTL1_);
#define UCB0BR0_             0x006A    /* USCI B0 Baud Rate 0 */
sfrb(UCB0BR0, UCB0BR0_);
#define UCB0BR1_             0x006B    /* USCI B0 Baud Rate 1 */
sfrb(UCB0BR1, UCB0BR1_);
#define UCB0I2CIE_           0x006C    /* USCI B0 I2C Interrupt Enable Register */
sfrb(UCB0I2CIE, UCB0I2CIE_);
#define UCB0STAT_            0x006D    /* USCI B0 Status Register */
sfrb(UCB0STAT, UCB0STAT_);
#define UCB0RXBUF_           0x006E    /* USCI B0 Receive Buffer */
const_sfrb(UCB0RXBUF, UCB0RXBUF_);
#define UCB0TXBUF_           0x006F    /* USCI B0 Transmit Buffer */
sfrb(UCB0TXBUF, UCB0TXBUF_);
#define UCB0I2COA_           0x0118    /* USCI B0 I2C Own Address */
sfrw(UCB0I2COA, UCB0I2COA_);
#define UCB0I2CSA_           0x011A    /* USCI B0 I2C Slave Address */
sfrw(UCB0I2CSA, UCB0I2CSA_);

// UART-Mode Bits
#define UCPEN                (0x80)    /* Async. Mode: Parity enable */
#define UCPAR                (0x40)    /* Async. Mode: Parity     0:odd / 1:even */
#define UCMSB                (0x20)    /* Async. Mode: MSB first  0:LSB / 1:MSB */
#define UC7BIT               (0x10)    /* Async. Mode: Data Bits  0:8-bits / 1:7-bits */
```

```
#define UCSPB              (0x08)     /* Async. Mode: Stop Bits  0:one / 1: two */
#define UCMODE1            (0x04)     /* Async. Mode: USCI Mode 1 */
#define UCMODE0            (0x02)     /* Async. Mode: USCI Mode 0 */
#define UCSYNC             (0x01)     /* Sync-Mode  0:UART-Mode / 1:SPI-Mode */

// SPI-Mode Bits
#define UCCKPH             (0x80)     /* Sync. Mode: Clock Phase */
#define UCCKPL             (0x40)     /* Sync. Mode: Clock Polarity */
#define UCMST              (0x08)     /* Sync. Mode: Master Select */

// I2C-Mode Bits
#define UCA10              (0x80)     /* 10-bit Address Mode */
#define UCSLA10            (0x40)     /* 10-bit Slave Address Mode */
#define UCMM               (0x20)     /* Multi-Master Environment */
//#define res               (0x10)     /* reserved */
#define UCMODE_0           (0x00)     /* Sync. Mode: USCI Mode: 0 */
#define UCMODE_1           (0x02)     /* Sync. Mode: USCI Mode: 1 */
#define UCMODE_2           (0x04)     /* Sync. Mode: USCI Mode: 2 */
#define UCMODE_3           (0x06)     /* Sync. Mode: USCI Mode: 3 */

// UART-Mode Bits
#define UCSSEL1            (0x80)     /* USCI 0 Clock Source Select 1 */
#define UCSSEL0            (0x40)     /* USCI 0 Clock Source Select 0 */
#define UCRXEIE            (0x20)     /* RX Error interrupt enable */
#define UCBRKIE            (0x10)     /* Break interrupt enable */
#define UCDORM             (0x08)     /* Dormant (Sleep) Mode */
#define UCTXADDR           (0x04)     /* Send next Data as Address */
#define UCTXBRK            (0x02)     /* Send next Data as Break */
#define UCSWRST            (0x01)     /* USCI Software Reset */

// SPI-Mode Bits
//#define res               (0x20)     /* reserved */
//#define res               (0x10)     /* reserved */
//#define res               (0x08)     /* reserved */
//#define res               (0x04)     /* reserved */
//#define res               (0x02)     /* reserved */

// I2C-Mode Bits
//#define res               (0x20)     /* reserved */
#define UCTR               (0x10)     /* Transmit/Receive Select/Flag */
#define UCTXNACK           (0x08)     /* Transmit NACK */
#define UCTXSTP            (0x04)     /* Transmit STOP */
#define UCTXSTT            (0x02)     /* Transmit START */
#define UCSSEL_0           (0x00)     /* USCI 0 Clock Source: 0 */
#define UCSSEL_1           (0x40)     /* USCI 0 Clock Source: 1 */
#define UCSSEL_2           (0x80)     /* USCI 0 Clock Source: 2 */
#define UCSSEL_3           (0xC0)     /* USCI 0 Clock Source: 3 */

#define UCBRF3             (0x80)     /* USCI First Stage Modulation Select 3 */
#define UCBRF2             (0x40)     /* USCI First Stage Modulation Select 2 */
#define UCBRF1             (0x20)     /* USCI First Stage Modulation Select 1 */
#define UCBRF0             (0x10)     /* USCI First Stage Modulation Select 0 */
#define UCBRS2             (0x08)     /* USCI Second Stage Modulation Select 2 */
#define UCBRS1             (0x04)     /* USCI Second Stage Modulation Select 1 */
#define UCBRS0             (0x02)     /* USCI Second Stage Modulation Select 0 */
#define UCOS16             (0x01)     /* USCI 16-times Oversampling enable */

#define UCBRF_0            (0x00)     /* USCI First Stage Modulation: 0 */
#define UCBRF_1            (0x10)     /* USCI First Stage Modulation: 1 */
#define UCBRF_2            (0x20)     /* USCI First Stage Modulation: 2 */
#define UCBRF_3            (0x30)     /* USCI First Stage Modulation: 3 */
#define UCBRF_4            (0x40)     /* USCI First Stage Modulation: 4 */
#define UCBRF_5            (0x50)     /* USCI First Stage Modulation: 5 */
#define UCBRF_6            (0x60)     /* USCI First Stage Modulation: 6 */
#define UCBRF_7            (0x70)     /* USCI First Stage Modulation: 7 */
#define UCBRF_8            (0x80)     /* USCI First Stage Modulation: 8 */
#define UCBRF_9            (0x90)     /* USCI First Stage Modulation: 9 */
#define UCBRF_10           (0xA0)     /* USCI First Stage Modulation: A */
#define UCBRF_11           (0xB0)     /* USCI First Stage Modulation: B */
```

```
#define UCBRF_12            (0xC0)      /* USCI First Stage Modulation: C */
#define UCBRF_13            (0xD0)      /* USCI First Stage Modulation: D */
#define UCBRF_14            (0xE0)      /* USCI First Stage Modulation: E */
#define UCBRF_15            (0xF0)      /* USCI First Stage Modulation: F */

#define UCBRS_0             (0x00)      /* USCI Second Stage Modulation: 0 */
#define UCBRS_1             (0x02)      /* USCI Second Stage Modulation: 1 */
#define UCBRS_2             (0x04)      /* USCI Second Stage Modulation: 2 */
#define UCBRS_3             (0x06)      /* USCI Second Stage Modulation: 3 */
#define UCBRS_4             (0x08)      /* USCI Second Stage Modulation: 4 */
#define UCBRS_5             (0x0A)      /* USCI Second Stage Modulation: 5 */
#define UCBRS_6             (0x0C)      /* USCI Second Stage Modulation: 6 */
#define UCBRS_7             (0x0E)      /* USCI Second Stage Modulation: 7 */

#define UCLISTEN            (0x80)      /* USCI Listen mode */
#define UCFE                (0x40)      /* USCI Frame Error Flag */
#define UCOE                (0x20)      /* USCI Overrun Error Flag */
#define UCPE                (0x10)      /* USCI Parity Error Flag */
#define UCBRK               (0x08)      /* USCI Break received */
#define UCRXERR             (0x04)      /* USCI RX Error Flag */
#define UCADDR              (0x02)      /* USCI Address received Flag */
#define UCBUSY              (0x01)      /* USCI Busy Flag */
#define UCIDLE              (0x02)      /* USCI Idle line detected Flag */

//#define res               (0x80)      /* reserved */
//#define res               (0x40)      /* reserved */
//#define res               (0x20)      /* reserved */
//#define res               (0x10)      /* reserved */
#define UCNACKIE            (0x08)      /* NACK Condition interrupt enable */
#define UCSTPIE             (0x04)      /* STOP Condition interrupt enable */
#define UCSTTIE             (0x02)      /* START Condition interrupt enable */
#define UCALIE              (0x01)      /* Arbitration Lost interrupt enable */


#define UCSCLLOW            (0x40)      /* SCL low */
#define UCGC                (0x20)      /* General Call address received Flag */
#define UCBBUSY             (0x10)      /* Bus Busy Flag */
#define UCNACKIFG           (0x08)      /* NAK Condition interrupt Flag */
#define UCSTPIFG            (0x04)      /* STOP Condition interrupt Flag */
#define UCSTTIFG            (0x02)      /* START Condition interrupt Flag */
#define UCALIFG             (0x01)      /* Arbitration Lost interrupt Flag */

#define UCIRTXPL5           (0x80)      /* IRDA Transmit Pulse Length 5 */
#define UCIRTXPL4           (0x40)      /* IRDA Transmit Pulse Length 4 */
#define UCIRTXPL3           (0x20)      /* IRDA Transmit Pulse Length 3 */
#define UCIRTXPL2           (0x10)      /* IRDA Transmit Pulse Length 2 */
#define UCIRTXPL1           (0x08)      /* IRDA Transmit Pulse Length 1 */
#define UCIRTXPL0           (0x04)      /* IRDA Transmit Pulse Length 0 */
#define UCIRTXCLK           (0x02)      /* IRDA Transmit Pulse Clock Select */
#define UCIREN              (0x01)      /* IRDA Encoder/Decoder enable */

#define UCIRRXFL5           (0x80)      /* IRDA Receive Filter Length 5 */
#define UCIRRXFL4           (0x40)      /* IRDA Receive Filter Length 4 */
#define UCIRRXFL3           (0x20)      /* IRDA Receive Filter Length 3 */
#define UCIRRXFL2           (0x10)      /* IRDA Receive Filter Length 2 */
#define UCIRRXFL1           (0x08)      /* IRDA Receive Filter Length 1 */
#define UCIRRXFL0           (0x04)      /* IRDA Receive Filter Length 0 */
#define UCIRRXPL            (0x02)      /* IRDA Receive Input Polarity */
#define UCIRRXFE            (0x01)      /* IRDA Receive Filter enable */

//#define res               (0x80)      /* reserved */
//#define res               (0x40)      /* reserved */
#define UCDELIM1            (0x20)      /* Break Sync Delimiter 1 */
#define UCDELIM0            (0x10)      /* Break Sync Delimiter 0 */
#define UCSTOE              (0x08)      /* Sync-Field Timeout error */
#define UCBTOE              (0x04)      /* Break Timeout error */
//#define res               (0x02)      /* reserved */
#define UCABDEN             (0x01)      /* Auto Baud Rate detect enable */
```

```
#define UCGCEN              (0x8000)  /* I2C General Call enable */
#define UCOA9               (0x0200)  /* I2C Own Address 9 */
#define UCOA8               (0x0100)  /* I2C Own Address 8 */
#define UCOA7               (0x0080)  /* I2C Own Address 7 */
#define UCOA6               (0x0040)  /* I2C Own Address 6 */
#define UCOA5               (0x0020)  /* I2C Own Address 5 */
#define UCOA4               (0x0010)  /* I2C Own Address 4 */
#define UCOA3               (0x0008)  /* I2C Own Address 3 */
#define UCOA2               (0x0004)  /* I2C Own Address 2 */
#define UCOA1               (0x0002)  /* I2C Own Address 1 */
#define UCOA0               (0x0001)  /* I2C Own Address 0 */

#define UCSA9               (0x0200)  /* I2C Slave Address 9 */
#define UCSA8               (0x0100)  /* I2C Slave Address 8 */
#define UCSA7               (0x0080)  /* I2C Slave Address 7 */
#define UCSA6               (0x0040)  /* I2C Slave Address 6 */
#define UCSA5               (0x0020)  /* I2C Slave Address 5 */
#define UCSA4               (0x0010)  /* I2C Slave Address 4 */
#define UCSA3               (0x0008)  /* I2C Slave Address 3 */
#define UCSA2               (0x0004)  /* I2C Slave Address 2 */
#define UCSA1               (0x0002)  /* I2C Slave Address 1 */
#define UCSA0               (0x0001)  /* I2C Slave Address 0 */

/************************************************************
* WATCHDOG TIMER
************************************************************/
#define __MSP430_HAS_WDT__            /* Definition to show that Module is available */

#define WDTCTL_            0x0120    /* Watchdog Timer Control */
sfrw(WDTCTL, WDTCTL_);
/* The bit names have been prefixed with "WDT" */
#define WDTIS0              (0x0001)
#define WDTIS1              (0x0002)
#define WDTSSEL             (0x0004)
#define WDTCNTCL            (0x0008)
#define WDTTMSEL            (0x0010)
#define WDTNMI              (0x0020)
#define WDTNMIES            (0x0040)
#define WDTHOLD             (0x0080)

#define WDTPW               (0x5A00)

/* WDT-interval times [1ms] coded with Bits 0-2 */
/* WDT is clocked by fSMCLK (assumed 1MHz) */
#define WDT_MDLY_32         (WDTPW+WDTTMSEL+WDTCNTCL)                        /* 32ms interval (default) */
#define WDT_MDLY_8          (WDTPW+WDTTMSEL+WDTCNTCL+WDTIS0)                 /* 8ms     " */
#define WDT_MDLY_0_5        (WDTPW+WDTTMSEL+WDTCNTCL+WDTIS1)                 /* 0.5ms   " */
#define WDT_MDLY_0_064      (WDTPW+WDTTMSEL+WDTCNTCL+WDTIS1+WDTIS0)          /* 0.064ms " */
/* WDT is clocked by fACLK (assumed 32KHz) */
#define WDT_ADLY_1000       (WDTPW+WDTTMSEL+WDTCNTCL+WDTSSEL)               /* 1000ms  " */
#define WDT_ADLY_250        (WDTPW+WDTTMSEL+WDTCNTCL+WDTSSEL+WDTIS0)        /* 250ms   " */
#define WDT_ADLY_16         (WDTPW+WDTTMSEL+WDTCNTCL+WDTSSEL+WDTIS1)        /* 16ms    " */
#define WDT_ADLY_1_9        (WDTPW+WDTTMSEL+WDTCNTCL+WDTSSEL+WDTIS1+WDTIS0) /* 1.9ms   " */
/* Watchdog mode -> reset after expired time */
/* WDT is clocked by fSMCLK (assumed 1MHz) */
#define WDT_MRST_32         (WDTPW+WDTCNTCL)                                /* 32ms interval (default) */
#define WDT_MRST_8          (WDTPW+WDTCNTCL+WDTIS0)                         /* 8ms     " */
#define WDT_MRST_0_5        (WDTPW+WDTCNTCL+WDTIS1)                         /* 0.5ms   " */
#define WDT_MRST_0_064      (WDTPW+WDTCNTCL+WDTIS1+WDTIS0)                  /* 0.064ms " */
/* WDT is clocked by fACLK (assumed 32KHz) */
#define WDT_ARST_1000       (WDTPW+WDTCNTCL+WDTSSEL)                        /* 1000ms  " */
#define WDT_ARST_250        (WDTPW+WDTCNTCL+WDTSSEL+WDTIS0)                 /* 250ms   " */
#define WDT_ARST_16         (WDTPW+WDTCNTCL+WDTSSEL+WDTIS1)                 /* 16ms    " */
#define WDT_ARST_1_9        (WDTPW+WDTCNTCL+WDTSSEL+WDTIS1+WDTIS0)          /* 1.9ms   " */

/* INTERRUPT CONTROL */
/* These two bits are defined in the Special Function Registers */
/* #define WDTIE              0x01 */
/* #define WDTIFG             0x01 */
```

```
/************************************************************
* Calibration Data in Info Mem
************************************************************/

#ifndef __DisableCalData

#define CALDCO_16MHZ_        0x10F8   /* DCOCTL   Calibration Data for 16MHz */
const_sfrb(CALDCO_16MHZ, CALDCO_16MHZ_);
#define CALBC1_16MHZ_        0x10F9   /* BCSCTL1 Calibration Data for 16MHz */
const_sfrb(CALBC1_16MHZ, CALBC1_16MHZ_);
#define CALDCO_12MHZ_        0x10FA   /* DCOCTL   Calibration Data for 12MHz */
const_sfrb(CALDCO_12MHZ, CALDCO_12MHZ_);
#define CALBC1_12MHZ_        0x10FB   /* BCSCTL1 Calibration Data for 12MHz */
const_sfrb(CALBC1_12MHZ, CALBC1_12MHZ_);
#define CALDCO_8MHZ_         0x10FC   /* DCOCTL   Calibration Data for 8MHz */
const_sfrb(CALDCO_8MHZ, CALDCO_8MHZ_);
#define CALBC1_8MHZ_         0x10FD   /* BCSCTL1 Calibration Data for 8MHz */
const_sfrb(CALBC1_8MHZ, CALBC1_8MHZ_);
#define CALDCO_1MHZ_         0x10FE   /* DCOCTL   Calibration Data for 1MHz */
const_sfrb(CALDCO_1MHZ, CALDCO_1MHZ_);
#define CALBC1_1MHZ_         0x10FF   /* BCSCTL1 Calibration Data for 1MHz */
const_sfrb(CALBC1_1MHZ, CALBC1_1MHZ_);

#endif /* #ifndef __DisableCalData */

/************************************************************
* Calibration Data in Info Mem
************************************************************/

/* TLV Calibration Data Structure */
#define TAG_DCO_30           (0x01)    /* Tag for DCO30  Calibration Data */
#define TAG_ADC10_1          (0x08)    /* Tag for ADC10_1 Calibration Data */
#define TAG_EMPTY            (0xFE)    /* Tag for Empty Data Field in Calibration Data */

#ifndef __DisableCalData
#define TLV_CHECKSUM_        0x10C0    /* TLV CHECK SUM */
const_sfrw(TLV_CHECKSUM, TLV_CHECKSUM_);
#define TLV_DCO_30_TAG_      0x10F6    /* TLV TAG_DCO30 TAG */
const_sfrb(TLV_DCO_30_TAG, TLV_DCO_30_TAG_);
#define TLV_DCO_30_LEN_      0x10F7    /* TLV TAG_DCO30 LEN */
const_sfrb(TLV_DCO_30_LEN, TLV_DCO_30_LEN_);
#define TLV_ADC10_1_TAG_     0x10DA    /* TLV ADC10_1 TAG */
const_sfrb(TLV_ADC10_1_TAG, TLV_ADC10_1_TAG_);
#define TLV_ADC10_1_LEN_     0x10DB    /* TLV ADC10_1 LEN */
const_sfrb(TLV_ADC10_1_LEN, TLV_ADC10_1_LEN_);
#endif

#define CAL_ADC_25T85        (0x0007)  /* Index for 2.5V/85Deg Cal. Value */
#define CAL_ADC_25T30        (0x0006)  /* Index for 2.5V/30Deg Cal. Value */
#define CAL_ADC_25VREF_FACTOR (0x0005) /* Index for 2.5V Ref. Factor */
#define CAL_ADC_15T85        (0x0004)  /* Index for 1.5V/85Deg Cal. Value */
#define CAL_ADC_15T30        (0x0003)  /* Index for 1.5V/30Deg Cal. Value */
#define CAL_ADC_15VREF_FACTOR (0x0002) /* Index for ADC 1.5V Ref. Factor */
#define CAL_ADC_OFFSET       (0x0001)  /* Index for ADC Offset */
#define CAL_ADC_GAIN_FACTOR  (0x0000)  /* Index for ADC Gain Factor */

#define CAL_DCO_16MHZ        (0x0000)  /* Index for DCOCTL  Calibration Data for 16MHz */
#define CAL_BC1_16MHZ        (0x0001)  /* Index for BCSCTL1 Calibration Data for 16MHz */
#define CAL_DCO_12MHZ        (0x0002)  /* Index for DCOCTL  Calibration Data for 12MHz */
#define CAL_BC1_12MHZ        (0x0003)  /* Index for BCSCTL1 Calibration Data for 12MHz */
#define CAL_DCO_8MHZ         (0x0004)  /* Index for DCOCTL  Calibration Data for 8MHz */
#define CAL_BC1_8MHZ         (0x0005)  /* Index for BCSCTL1 Calibration Data for 8MHz */
#define CAL_DCO_1MHZ         (0x0006)  /* Index for DCOCTL  Calibration Data for 1MHz */
#define CAL_BC1_1MHZ         (0x0007)  /* Index for BCSCTL1 Calibration Data for 1MHz */


/************************************************************
```

```
* Interrupt Vectors (offset from 0xFFE0)
************************************************************/

#define TRAPINT_VECTOR        (0x0000)  /* 0xFFE0 TRAPINT */
#define PORT1_VECTOR          (0x0004)  /* 0xFFE4 Port 1 */
#define PORT2_VECTOR          (0x0006)  /* 0xFFE6 Port 2 */
#define ADC10_VECTOR          (0x000A)  /* 0xFFEA ADC10 */
#define USCIAB0TX_VECTOR      (0x000C)  /* 0xFFEC USCI A0/B0 Transmit */
#define USCIAB0RX_VECTOR      (0x000E)  /* 0xFFEE USCI A0/B0 Receive */
#define TIMER0_A1_VECTOR      (0x0010)  /* 0xFFF0 Timer0)A CC1, TA0 */
#define TIMER0_A0_VECTOR      (0x0012)  /* 0xFFF2 Timer0_A CC0 */
#define WDT_VECTOR            (0x0014)  /* 0xFFF4 Watchdog Timer */
#define COMPARATORA_VECTOR    (0x0016)  /* 0xFFF6 Comparator A */
#define TIMER1_A1_VECTOR      (0x0018)  /* 0xFFF8 Timer1_A CC1-4, TA1 */
#define TIMER1_A0_VECTOR      (0x001A)  /* 0xFFFA Timer1_A CC0 */
#define NMI_VECTOR            (0x001C)  /* 0xFFFC Non-maskable */
#define RESET_VECTOR          (0x001E)  /* 0xFFFE Reset [Highest Priority] */


/************************************************************
* End of Modules
************************************************************/
#ifdef __cplusplus
}
#endif /* extern "C" */

#endif /* #ifndef __MSP430G2553 */
```

Appendix D
Debug in Code Composer

In the process of debugging, it is useful to see what is happening in the registers. It is possible to view the contents of registers from within the Code Composer IDE as you progress through the code listing line by line. The first step is to set up the tools necessary to accomplish this in Code Composer.

1. First thing you have to do is add "Assembly Step Into" and "Assembly Step Over" commands in the menu toolbar at the top. Find the "Quick Access" box on the right side of the menu bar and type in "Assembly Step Over." Once you see the commands appear, right click on them and they will appear on the menu bar as two green downward pointing arrows (one straight down and the other pointing down at an angle).
2. Next mark locations in the program you want to stop (the breakpoint) going through line by line. Most of the time this will be at the end of the program. Right click in the space left of the code listing numbers next to the line number where you want to stop. A box will appear and left click on "breakpoint." To undo the breakpoint right click then left click on "Toggle Breakpoint."
3. Next click on "View" then "Registers." A large box will appear on the right. To expand the choices of registers double click on the category (i.e. double click on Core Registers).
4. Debug the program as usual by clicking the "green bug" icon. Next do not run the program as usual, however, start stepping through the program line by line with the "Step Into" and "Step Over" buttons. Step Into takes you deep into the program and will look at things happening in the Header file (probably will not be necessary for what you may want to look at). Step Over will take you through the program line by line. As you use these two buttons you will see the values in the registers change.

   Note: Sometimes you may use a loop to create a delay (i.e. do a loop 65535 times for a 65.535 ms delay). If it is a short loop clicking through several times is not an issue, however, for a big loop, clicking through 65000+ times would not be practical. For purposes of debugging, change the number to a smaller value.

© The Editor(s) (if applicable) and The Author(s), under exclusive license to Springer Nature Switzerland AG 2023
J. Kretzschmar et al., *MSP430 Microcontroller Lab Manual*, Synthesis Lectures on Digital Circuits & Systems, https://doi.org/10.1007/978-3-031-26643-0

Appendix E
Controlling a 16×2 LCD Display

E.1   Introduction

This appendix describes how to interface a common 16×2 LCD Display to the Texas Instruments MSP430G2553 microcontroller.[1] Datasheets are difficult to decipher sometimes so this descriptive explanation should ease the process. There are many 16×2 LCD displays available and the one chosen to use is the Newhaven NHD–0216K1Z (www. newhavendisplay.com, price is about US$13). This display accepts parallel data input and has white characters with a blue backlight which makes it very easy to see as shown in Fig. E.1.

The code was written to drive the display in the 8-bit data delivery mode. From an efficiency standpoint this is not ideal because more pins are used on the microcontroller, however when learning, the 8–bit mode is easier to setup, follow, and understand. There are delay routines using loops, this also is not efficient use of program memory space, however, it is easier to understand. The delays used in the LCD_PORT_2 code listing are longer than described in the datasheet so there is ample time for the LCD display to process the instructions and data. In the code listing the delays are based upon the CPU clock set at 16 MHz, and the timing was determined by viewing on an oscilloscope.

---

[1] This article originally appeared in "Controlling a 16x2 LCD with the Texas Instruments MSP430G2553 Microcontroller," QEX–A Forum for Communications Experimenters, No.199 Nov/Dec 2021. We gratefully acknowledge the American Radio Relay League (ARRL), www.arrl. org, for permission to include it in this book.

© The Editor(s) (if applicable) and The Author(s), under exclusive license to Springer   159
Nature Switzerland AG 2023
J. Kretzschmar et al., *MSP430 Microcontroller Lab Manual*, Synthesis Lectures on Digital
Circuits & Systems, https://doi.org/10.1007/978-3-031-26643-0

Fig. E.1 16×2 LCD display with the MSP430G2553 microcontroller

E.2   Interfacing the MSP430G2553 and the NHD–0216K1Z

The LCD has 16 solder pads labeled #1 through #16 (on both sides of the module). Choose the side that works the best for your project and solder the header pins on that side. On the right side of the module there are two solder pads. One is labeled as "K1" and this is for "Ground" and the other is labeled as "A1" and this is for "+5 V." These provide power for the backlight. Reference Fig. E.2.

E.3   Sending Instructions to the LCD Display

The LCD display needs a specific sequence of instructions delivered to it to make it function. First, the Wake–up sequence which involves sending special function COMMANDS. Second, a sequence of COMMANDS is sent into the LCD display instructing it what to do (such as: Clear Screen, Return to Home, etc.). These COMMANDS are described in detail in the Configure the LCD Display section.

The one time "wake–up sequence" involves a minimum of a 40 ms delay after power is applied to the LCD display, and sending a special function COMMAND two times (three times when using the HD44780 LCD driver chip). Sending the command instruction #00111000b (Hex 0x38) for the two (or three) times that it is required will set up the LCD display for 8–bit format, and a two line display. The wake–up sequence code in LCD_PORT_2 sends the special function COMMAND three times to cover both the ST7066U and HD44780 driver chip requirements. The extra instruction information does not affect the operation of the LCD display.

The LCD display needs to be informed as to what type of information it will be receiving. Pin #4 of the LCD display is the Register Select (R/S) pin. Placing a "0" on Pin #4 tells the LCD display to get ready because the next information you will receive will be a COMMAND

Ground	Pin #1	WIRED CONNECTION	Pin #20	(Ground)
+5.0 Volts	Pin #2	See Notes Below		
Contrast Control	Pin #3	See Notes Below		
R/S Register Select	Pin #4	WIRED CONNECTION	Pin #6	(P1.4)
R/W Read - Write	Pin #5	See Notes Below		
Enable	Pin #6	WIRED CONNECTION	Pin #7	(P1.5)
Data-0	Pin #7	WIRED CONNECTION	Pin #8	(P2.0)
Data-1	Pin #8	WIRED CONNECTION	Pin #9	(P2.1)
Data-2	Pin #9	WIRED CONNECTION	Pin #10	(P2.2)
Data-3	Pin #10	WIRED CONNECTION	Pin #11	(P2.3)
Data-4	Pin #11	WIRED CONNECTION	Pin #12	(P2.4)
Data-5	Pin #12	WIRED CONNECTION	Pin #13	(P2.5)
Data-6	Pin #13	WIRED CONNECTION	Pin #19	(P2.6)
Data-7	Pin #14	WIRED CONNECTION	Pin #18	(P2.7)
Backlight (+5.0 v)	Pin #15			
Backlight (Ground)	Pin #16			

LCD DISPLAY   **MSP430G2553**

Notes:
- Pin #2 and Pin #15 - +5.0 volts (Use MSP430G2553 +5.0 volt supply)
- Pin #3 - Connect to wiper (middle terminal of a 10K pot)
- Pin #5 - Ground for writing to LCD Display

Fig. E.2 16×2 LCD display with the MSP430G2553 connections

(such as the special function COMMAND). Placing a "1" on Pin #4 tells the LCD display to get ready because the next information received will be DATA (such as DATA from a sensor).

When sending a COMMAND or DATA, first the Enable pin, Pin #6 on the LCD display must be brought High ("1") and kept there for a short period of time. The COMMAND or DATA information is then placed on the data Pins #7–14. The Enable Pin #6 is then brought Low ("0") and kept there for a short period of time. The delay time for keeping the

Enable pin in the two states is not specified in the datasheet so 5 ms was selected. Regardless of whether COMMAND information or DATA information is sent to the LCD display the following steps have to happen:

1. COMMAND or DATA information designated by LCD display Pin #4
2. LCD Pin #6 (Enable) needs to be brought "High" ("1") (Delay program code: 5 ms)
3. Information (COMMAND or DATA) is placed on Pins #7–14
4. LCD Pin #6 (Enable) needs to be brought "Low" ("0") (Delay program code: 5 ms)

E.4   Configuring the LCD Display

Wake-up Sequence:

1. Power applied to LCD display (Delay more than 40 ms, program code: 50 ms)
2. Special function COMMAND (00111000b or 0x38) (Delay of more than 4.1 ms, program code: 5 ms)
3. Special function COMMAND (00111000b or 0x38) (Delay of more than 100 μs, program code: 5 ms)
4. Special function COMMAND (00111000b or 0x38) (Delay of more than 100 μs, program code: 5 ms)

Now that the LCD display has been "woken up" the next steps are to configure the display for a particular need. COMMANDS can be sent to the LCD display setting up such parameters as where to display characters or which direction the next characters will be displayed.

The table of commands (Fig. E.3) shows the COMMAND "base" number (the bit position for where a "1" is present). Five of the COMMANDS allow you to select options (where an "X" appears in the binary number). When options are selected, a "0" or a "1" is placed in the bit position and added to the "base" number, this new number specifies a particular option.

For simplicity only five COMMANDS will be discussed in detail: (1) Clear Display, (2) Display Control, (3) Display Shift, (4) Entry Mode, and (5) Setting of the Character Position Address (DDRAM address). The three remaining COMMANDS involve advanced display options, and special character generation and can be investigated by reviewing the datasheet if needed. Each COMMAND will be detailed with pictures showing how the LCD display responded.

The five commands will be discussed in the following order: (1) Selection of 8–Bit/2 Line mode, (2) Clear Display, (3) Setting of the DDRAM address where you want characters to be displayed, (4) Display Control, and (5) Display Shift (left or right).

COMMAND	BASE CODE	WHAT HAPPENS
Clear Display	00000001b	Display is cleared
Cursor Home	0000001*b	Cursor goes home (00)
Entry Mode	000001XXb	How characters displayed
Display Control	00001XXXb	On/Off, Cursor, Blink
Cursor Display Shift	000100**b	Cursor Shift control
8-Bit or 4-Bit Setup	001XXX**b	8-Bit/4-Bit, 1 or 2 lines
CGRAM Setup	01XXXXXXb	Character Generator
Sets DDRAM Address	1XXXXXXXb	Sets DDRAM Address

* = Bits that are ignored (in the program code just make them a "0")
X = Bits that an option needs to be selected

Fig. E.3 LCD commands

E.4.1   Commands and Examples

Command: Selecting 8–Bit/4–Bit mode, 1 or 2 Lines (001XXX**b) (Fig. E.4)

- BIT–0: Ignored (make = 0)
- BIT–1: Ignored (make = 0)
- BIT–2: 0–Font of characters is 5x7 Dots (always select this option)
- BIT–2: 1–Font of characters is 5x10 Dots (not common)
- BIT–3: 0–Information displayed on 1 line
- BIT–3: 1–Information displayed on 2 lines (always select this option)
- BIT–4: 0–Commands and Data sent in 4–Bit format
- BIT–4: 1–Commands and Data sent in 8–Bit format

Example: #00111000b Selects 8–bit mode/2 line mode for entry of information

Fig. E.4 LCD selecting 8/4 bit
mode. Note no discernible
change

Fig. E.5 Clears Display and
places cursor at home position

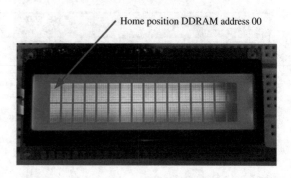

Home position DDRAM address 00

Command: Clear Display (00000001b)

Example: #00000001b Clears Display and places cursor at home position (Fig. E.5)

Command: Sets DDRAM Address (1XXXXXXXb) This instruction sets the position on the 16×2 LCD display where the first character will be displayed. There are 32 possible DDRAM addresses (a specific number for each location). Arriving at the correct number to enter can be confusing. The datasheet gives a number, however, that number needs to be added to the base number #10000000b (Fig. E.6).

Example: Letter "K" placed at DDRAM address 73 (Fig. E.7).

$$\#10000000b \ = \ (base\ number\ 128)$$
$$\#01001001b \ = \ (73)$$

$$128 \ + \ 73 \ = \ 201$$

Conversion to Binary: #11001001b = 201

#11001001b This value places a "K" in DDRAM Address 73.

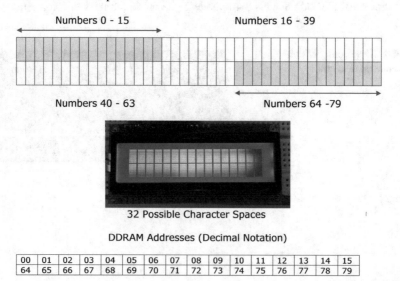

Numbers 0 - 15

Numbers 16 - 39

Numbers 40 - 63

Numbers 64 -79

32 Possible Character Spaces

DDRAM Addresses (Decimal Notation)

00	01	02	03	04	05	06	07	08	09	10	11	12	13	14	15
64	65	66	67	68	69	70	71	72	73	74	75	76	77	78	79

Fig. E.6 DDRAM address positions

Fig. E.7 Letter "K" placed at DDRAM address 73

Command: Display Control (00001XXXb) (Fig. E.8)

- BIT–0: 0–Cursor Blinking OFF
- BIT–0: 1–Cursor Blinking ON
- BIT–1: 0–Cursor OFF
- BIT–1: 1–Cursor ON
- BIT–2: 0–Display OFF
- BIT–2: 1–Display ON

#00001000b   Display OFF, Cursor OFF, Cursor Blinking OFF

#00001100b   Display ON, Cursor OFF, Cursor Blinking OFF

#00001110b   Display ON, Cursor ON, Cursor Blinking OFF

#00001111b   Display ON, Cursor ON, Cursor Blinking ON

Fig. E.8 LCD display control

Command: Display Shift/Cursor Move (Right or Left) (000001XXb) (Fig. E.9)

- BIT–0: 0–Display shift OFF
- BIT–0: 1–Display shift ON
- BIT–1: 0–Cursor moves one space to the left (for next character)
- BIT–1: 1–Cursor moves one space to the right (for next character)

#00000100b   Display Shift OFF / Cursor to the Left

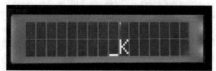

(The K stays in DDRAM 73 while the next characters go in on the Left)

#00000110b   Display Shift OFF / Cursor to the Right

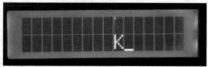

(The K stays in DDRAM 73 while the next characters go in on the right)

#00000101b   Display Shift ON / Cursor to the Left

(The Cursor stays in DDRAM 73 and stays on the Left while the K moves)

#00000111b   Display Shift ON / Cursor to the Right

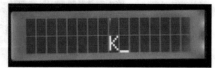

(The Cursor stays in DDRAM 73 and stays on the Right while the K moves)

Fig. E.9 LCD display cursor move

Appendix F
LCD Display Using C with the MSP430G2553

F.1   Overview

The attached code listing was developed to interface the MSP430G2553 microcontroller with a 16×2 LCD Display. The display used is the Newhaven NHD–0216K1Z. For demonstration purposes the code engages one channel of the Analog–to–Digital Converter (ADC) connected to a 5K potentiometer. The potentiometer operates as a voltage divider and inputs a voltage between 0 V and 3.3 V into pin P1.7 of the MSP430G2553. This voltage is then converted into a number between 0 and 1023 and sent to the LCD display. The potentiometer functions as a rotational sensor; however, other types of sensors could be used in place of the potentiometer. Detailed explanations of the C code are provided in the code listing.[2]

Fig. F.1 LCD display with ADC output

---

[2] This appendix was written by D. Kretzschmar.

J. Kretzschmar et al., *MSP430 Microcontroller Lab Manual*, Synthesis Lectures on Digital Circuits & Systems, https://doi.org/10.1007/978-3-031-26643-0

F.2   Hardware Connections

The hardware connections for this laboratory demonstration are provided in Fig. F.2.

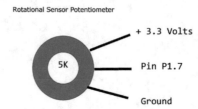

LCD Display Connections to MSP430G2553

To be used with Program: C_CODE_ADC_LCD_PORT_2

LCD DISPLAY			MSP430G2553	
Ground	Pin #1	------------→	Pin #20	(Ground)
+5.0 Volts	Pin #2	See NOTE		
Contrast Control	Pin #3	See NOTE		
R/S Register Select	Pin #4	------------→	Pin #6	(P1.4)
R/W Read - Write	Pin #5	------------→	Pin #4	(P1.2)
Enable	Pin #6	------------→	Pin #7	(P1.5)
Data-0	Pin #7	------------→	Pin #8	(P2.0)
Data-1	Pin #8	------------→	Pin #9	(P2.1)
Data-2	Pin #9	------------→	Pin #10	(P2.2)
Data-3	Pin #10	------------→	Pin #11	(P2.3)
Data-4	Pin #11	------------→	Pin #12	(P2.4)
Data-5	Pin #12	------------→	Pin #13	(P2.5)
Data-6	Pin #13	------------→	Pin #19	(P2.6)
Data-7	Pin #14	------------→	Pin #18	(P2.7)
Backlight (+5.0 v)	Pin #15	See NOTE		
Backlight (Ground)	Pin #16	------------→	Pin #20	(Ground)

NOTE

```
LCD Pin #2  -- +5.0 volts (Use Launchpad +5.0 volt supply)
LCD Pin #15 -- +5.0 volts (Use Launchpad +5.0 volt supply)
LCD Pin #3  -- Connect to middle terminal of a 10K potentiometer
LCD Pin #5  -- For LCD Write/Read operations connect to P1.2
```

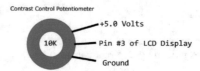

Fig. F.2 Hardware connections

F.3 LCD Display C Code

```
/*******************************************************************************
*                                  main.c
*
* DISCLAIMER:  This C Language program was designed and intended as additional
*              material to help introduce basic applications using the
*              MSP430 microcontroller. Detailed explanations are provided with
*              the actual code to help understand program processes. This code
*              is not optimized for efficiency, or speed and is strictly
*              designed for learning purposes. The code can be modified and
*              used for other applications. It is the responsibility of the
*              end-user to ensure particular applications have been thoroughly
*              tested.
********************************************************************************
* Name:        EXAMPLE_LCD_ADC
*
* Description: C Code implementation of Assembly program ADC_LCD_PORT_2.
*
*              - Uses ADC module on the MSP430G2553 and displays the output
*                number on a 16x2 LCD display
*              - Checks LCD display Busy Flag
*              - Checks ADC Busy Flag
*              - Accommodates data values up to 5 digits for future projects
*              - Addition and subtraction operations were used instead
*                of the arithmetic operators (*/%) to help the reader
*                understand the logic behind multiplication and division.
*                Bitwise operations could have also been utilized. For details
*                and code examples with bitwise operators, see TI SLAA329A
*                ''Efficient Multiplication and Division Using MSP430 MCUs''.
*
* Parts:       - Texas Instruments MSP-EXP430G2ET Launchpad with MSP430G2553
*              - Newhaven NHD-0216K1Z 16x2 LCD Display (or similar display)
*              - 10K potentiometer (for contrast control)
*              - 5K potentiometer (for voltage divider for ADC input)
*
* References:  - Texas Instruments SLAU144K Users Guide (www.ti.com)
*              - Texas Instruments SLAS735J MSP430G2553 datasheet (www.ti.com)
*              - Texas Instruments MSP430 Optimizing C/C++ Compiler Users
*                Guide, v21.6.0.LTS (www.ti.com). For detailed
*                explanations of the data types and their size used in this program
*                that are specific for the MSP430 architecture, see
*                chapter 5.5 ''Data Types''.
*              - Newhaven NHD-0216K1Z datasheet (www.newhavendisplay.com)
*              - Paper "Controlling a 16x2 LCD Display with the MSP430G2553
*                Microcontroller Using Assembly Code" (Appendix)
*              - Paper "Interfacing the MSP430G2553 Microcontroller with an
*                LCD Display Using C Code" (Appendix)
*
* Connections:
*
*    LCD Display      MSP430G2553
*      Pin #1           Ground
*      Pin #2           +5 volts
*      Pin #3           Contrast control (center pin from 10K potentiometer)
*      Pin #4           P1.4 (R/S Register Select)
*      Pin #5           P1.2 (Read/Write Select)
*      Pin #6           P1.5 (Enable)
*      Pin #7           P2.0 (Data)
*      Pin #8           P2.1 (Data)
```

```
*       Pin #9          P2.2 (Data)
*       Pin #10         P2.3 (Data)
*       Pin #11         P2.4 (Data)
*       Pin #12         P2.5 (Data)
*       Pin #13         P2.6 (Data)
*       Pin #14         P2.7 (Data)
*       Pin #15         +5 volts (backlight)
*       Pin #16         Ground (backlight)
*
*                       P1.7 (center pin from 5K potentiometer)
*                       Ground (one end of potentiometer)
*                       +3.3 volts (one end of potentiometer)
*
*
* Notes:          Additional explanation regarding the code
*
*       Note 1: When the code is compiled in Code Composer, there may be
*               remarks stating that using a timer module is recommended
*               instead of __delay_cycles()
*
*               __delay_cycles() is an intrinsic function that is accurate and
*               easy to use. Implementing a timer is more advanced.
*
*       Note 2: #define SIZE 6 sets the size of a character (char) array.
*               This array is used to create a C-style string or C-String that
*               holds the ASCII values of the ADC OUTPUT once it is converted
*               to ASCII values. This information is then displayed on the LCD.
*               Depending on the maximum number of digits in the data, the
*               size of the array can be varied.
*
*               IMPORTANT: in order for an array of chars (characters) to be a
*               C-style string or C-String, it must be NULL terminated. The
*               NULL terminator NULL or '\0' lets the compiler know where the
*               string ends.
*
*               For example, when the ADC outputs values from 0 to 1023 then
*               the char array needs a minimum SIZE 5 to accommodate 4 digits.
*               Four of the indices of the array are used for the output and
*               the 5th index is used for the NULL terminator of the C-String.
*
*               SIZE 6 was used in this program to accommodate projects
*               in the lab manual that display values with up to 5 digits.
*
*       Note 3: Positioning the Cursor on the LCD display
*
*               The LCD display has 2 rows, each with 16 columns. The rows
*               are numbered 0 and 1 and the columns are numbered 0 - 15.
*
*           Use the table as a guide for positioning the first character.
*
*               R = Row and C = Column
*
*    | C0| C1| C2| C3| C4| C5| C6| C7| C8| C9| C10| C11| C12| C13| C14| C15
*   R0|0,0|0,1|0,2|0,3|0,4|0,5|0,6|0,7|0,8|0,9|0,10|0,11|0,12|0,13|0,14|0,15
*   R1|1,0|1,1|1,2|1,3|1,4|1,5|1,6|1,7|1,8|1,9|1,10|1,11|1,12|1,13|1,14|1,15
*
*               Example:
*                       To start the display on the first row and 4th column,
*                       the following function call would be used.
```

```
 *                            lcd_initialize (0,3);
 *
 *      Note 4: Checking the busy flag in the ADC
 *
 *              The ADC needs time to complete the conversion process. Bit 0
 *              of register ADC10CTL1 is the ADC10BUSY flag. However, if
 *              the statement "while(ADC10CTL1 & ADC10BUSY);" is used, the
 *              optimizer will output the following recommendation:
 *
 *              "Detected flag polling using ADC10BUSY. Recommend using an
 *              interrupt combined with enter LPMx and ISR"
 *
 *              An alternative way to check the ADC busy flag status is to
 *              follow the example in the SLAU144K User's Guide. This example
 *              uses the fact that BUSY is defined in the header file as
 *              0x0001. ADC10BUSY is also defined in the header file as 0x0001.
 *              When the statement "while(ADC10CTL1 & BUSY);" is used the
 *              while loop continues until Bit 0 of ADC10CTL1 is zero, and
 *              the optimizer does not print any recommendations.
 *
 *      Note 5: Delay at bottom of while(1) loop
 *
 *              A 500 millisecond delay is used before restarting the loop. This
 *              minimizes flicker that occurs as data is displayed on the LCD.
 *              This delay can be adjusted as needed. A detailed explanation
 *              of how to calculate the number of clock cycles to delay for a
 *              specified time can be found in msp430_lcd_adc.c
 *
 * Author:      D. Kretzschmar
 *
 * Updated:     December 3, 2022
 ******************************************************************************/

#include <msp430.h>
#include "msp430_lcd_adc.h"


 /*  #define SIZE 6 sets the size of the character (char) array used in this
  *  program. SIZE may be changed by the user. Please see Note 2 in the header
  *  of this file for details*/
#define SIZE 6


int main(void)
{
    WDTCTL = WDTPW | WDTHOLD;    /* Stop the watchdog timer*/

    /* Set the clock speed to 16 Megahertz */
    BCSCTL1 = CALBC1_16MHZ;
    DCOCTL = CALDCO_16MHZ;

    /* Variables to hold row and column information for LCD display.
     * Please see Note 3 for details/table on how to determine integer values
     * to use. */
    unsigned int row = 0;
    unsigned int column = 3;

    lcd_initialize(row, column); /* Initialize the LCD display */
```

```c
ADC_initialize(); /* Call function that initializes ADC */

unsigned int ADC_OUT; /* Variable to hold the ADC output */

/* Declare array of chars to hold ADC OUTPUT after conversion to ASCII
 * values. Indices initialized to NULL */
char string_with_ADC_OUTPUT[SIZE] = { NULL };

/* Display characters (up to 16) in between quotations.
 * Characters will display starting at row/column initialized above. */
display_info("ADC OUTPUT");


while (1) /* Never-ending loop */
{
    ADC10CTL0 = ADC10CTL0 | (ENC | ADC10SC); /*Enables + starts conversion*/

    while (ADC10CTL1 & BUSY); /* Check if ADC busy -- Please see NOTE 4 in
                               * file header for detailed explanation */

    ADC_OUT = ADC10MEM; /* Transfer the data from ADC to variable ADC_OUT */

    /* Convert the unsigned integer value in ADC_OUT to ASCII values */
    integer_to_ASCII(string_with_ADC_OUTPUT, ADC_OUT, SIZE);

    /* Adjust the row/column where the ADC OUTPUT will display to avoid
     * overwriting the label "ADC OUTPUT" (currently positioned at
     * at row = 0, column = 3). */
    row = 1;
    column = 6;

    /* Reset where the LCD cursor will start writing */
    set_DDRAM_address(row, column);


    display_info(string_with_ADC_OUTPUT); /* Display data on LCD */

    /* Pause for 500 milliseconds. See Note 5 for details on why this
     * length of delay was chosen. */
    __delay_cycles(8000000);

}

}


/***************************************************************************/


/***************************************************************************
 *                     msp430_lcd_adc.h
 *
 * DISCLAIMER:  This C Language program was designed and intended as additional
 *              material to help introduce basic applications using the
 *              MSP430 microcontroller. Detailed explanations are provided with
 *              the actual code to help understand program processes. This code
 *              is not optimized for efficiency, or speed and is strictly
 *              designed for learning purposes. The code can be modified and
 *              used for other applications. It is the responsibility of the
 *              end-user to ensure particular applications have been thoroughly
 *              tested.
```

```
*********************************************************************************
* Description: This file contains the MACRO definitions and function
*              declarations.
*
*              The function declarations are in order of their definitions
*              found in msp430_lcd_adc.c. The functions were ordered by their
*              use in the following manner:
*              1. Functions relating to the LCD
*              2. Function relating to the ADC
*              3. Utility functions such as mathematical operations
*
*  Author:     D. Kretzschmar
*
*  Updated:    December 3, 2022
*********************************************************************************/

#ifndef MSP430_LCD_ADC_H_
#define MSP430_LCD_ADC_H_

#include <msp430.h>
#include <stdio.h>

/*******************************************************************************
 * DEFINE THE THREE CONTROL PINS THAT SEND OUT CONTROL SIGNALS
 *
 *     (1) R/W --- READ/WRITE  MSP430G2553 pin P1.2
 *
 *     (2) RS --- Register Select MSP430G2553 pin P1.4
 *         COMMAND_MODE = 0
 *         DATA_MODE = 1
 *
 *     (3) ENABLE --- MSP430G2553 pin P1.5
 *
 *         This program follows the reading/writing timing diagrams in the
 *         ST7066U datasheet.
 *******************************************************************************/
#define READ_MODE P1OUT = P1OUT | BIT2 /* Setting R/W Pin P1.2 to  HIGH (1) */
#define WRITE_MODE P1OUT = P1OUT & ~(BIT2)/* Setting R/W Pin P1.2 to LOW (0)*/

#define DATA_MODE P1OUT = P1OUT | BIT4 /* Setting RS Pin P1.4 to HIGH (1) */
#define COMMAND_MODE P1OUT = P1OUT & (~BIT4) /*Setting RS Pin P1.4 to LOW (0)*/

#define ENABLE_PIN_HIGH P1OUT = P1OUT | BIT5   /* setting Pin P1.5 to HIGH */
#define ENABLE_PIN_LOW P1OUT = P1OUT & (~BIT5) /* setting Pin P1.5 to LOW */

/*******************************************************************************
  * FUNCTION DECLARATIONS RELATING TO THE LCD
  *******************************************************************************/
void lcd_initialize(unsigned int row, unsigned int column);
void initialize_ports();
void check_busy_flag();
void set_DDRAM_address(unsigned int row, unsigned int column);
void send_instruction_to_lcd(unsigned int instruction, unsigned int check_busy);
void write_ASCII_to_lcd(char ASCII);
void display_info(char* information);

/*******************************************************************************
  * FUNCTION DECLARATIONS RELATING TO THE ADC
  *******************************************************************************/
```

```
void ADC_initialize();

/*****************************************************************************
 * UTILITY FUNCTIONS
 *****************************************************************************/
const char* integer_to_ASCII(char ascii[], unsigned int value, unsigned int length);
const char* clearString(char string[], unsigned int length);
unsigned int number_of_digits(unsigned int value);
unsigned int division(unsigned int dividend, unsigned int divisor);
unsigned int get_modulo(unsigned int dividend, unsigned int divisor);
unsigned int multiply(unsigned int multiplicand, unsigned int multiplier);

#endif /* MSP430_LCD_ADC_H_ */

/*****************************************************************************/


/*****************************************************************************
 *                          msp430_lcd_adc.c
 *
 * DISCLAIMER:  This C Language program was designed and intended as additional
 *              material to help introduce basic applications using the
 *              MSP430 microcontroller. Detailed explanations are provided with
 *              the actual code to help understand program processes. This code
 *              is not optimized for efficiency, or speed and is strictly
 *              designed for learning purposes. The code can be modified and
 *              used for other applications. It is the responsibility of the
 *              end-user to ensure particular applications have been thoroughly
 *              tested.
 *****************************************************************************
 * Description: Contains explanation on how to use __delay_cyles() and function
 *              definitions.
 *
 *              - The function definitions appear in the order declared in the
 *                header file.
 *              - Each function definition contains a description of the
 *                purpose, the parameters, return (if any), and any applicable
 *                references.
 *              - Additional Notes with detailed information
 *
 * Author:      D. Kretzschmar
 *
 * Updated:     December 3, 2022
 *****************************************************************************/

#include <msp430_lcd_adc.h>

/*****************************************************************************
 *                          __delay_cycles(X)
 *
 * Name:        void __delay_cycles(unsigned long X)
 *
 * Description: Per the manual referenced below, the __delay_cycles() intrinsic
 *              function inserts code that delays the program by the number of
 *              of clock cycles indicated by X (shown above).
 *              The number of cycles must be a constant i.e., a number such
 *              as 100000 or a defined constant value.
 *
 * Examples:    Below are 2 examples that show how to calculate the number of
 *              clock cycles needed for "X" to result in a delay for the
```

```
*              desired length of time.
*
*              EXAMPLE 1:
*
*              With the CPU clock set at 1 MHz (1000000 cycles/sec);
*              for a delay of 0.050 sec (50 milliseconds) use the
*              calculation below.
*
*              CALCULATIONS: Set up a ratio
*              1.0 sec/1000000 clock cycles/sec = 0.05 sec/X clock cycles/sec
*              X = (1000000 x 0.05) clock cycles
*              X = 50000 clock cycles
*              Here is how to code the delay: __delay_cycles(50000);
*
*              EXAMPLE 2:
*
*              With the CPU clock set at 16 MHz (16000000 cycles/sec);
*              for a delay of 0.050 sec (50 milliseconds) use the
*              calculation below.
*
*              CALCULATIONS: Set up a ratio
*              1.0 sec/16000000 clock cycles/sec = 0.05 sec/X clock cycles/sec
*              X = (16000000 x 0.05) clock cycles
*              X = 800000 clock cycles
*              Here is how to code the delay: __delay_cycles(800000);
*
* Parameter:  Takes an unsigned long integer which represents the number of
*             clock cycles to delay
*
* Return:     None
*
* Reference:  MSP430 Optimizing C/C++ Compiler, v21.6.0.LTS, User's Guide
*             found at: (www.ti.com)
***************************************************************************/

                          /*FUNCTION DEFINITIONS */

/***************************************************************************
* Name:      lcd_initialize(unsigned int row, unsigned int column)
*
* Description: Purpose is to designate I/O pins and designate which pins
*              represent the OUTPUT pins.
*
*              - Function also allows the user to select the DDRAM address by
*                selecting the row and column where the LCD cursor initially
*                starts
*              - The remainder of the setup has default settings. The display
*                shift is on, the cursor moves to the right, and does not
*                blink
*              - The more advanced user can look at the paper in the Appendix:
*                "Controlling a 16x2 LCD Display with the MSP430G2553
*                Microcontroller Using Assembly Code"
*
* Parameters: Two unsigned integers. The first represents the row that the
*             display will start at, and the second unsigned integer
*             represents the column. The LCD has 2 rows and 16 columns. The
*             rows are numbered 0 and 1 and the columns are numbered 0 - 15.
*
*             Use the table as a guide for positioning the first character.
```

```
*
*                    R = Row and C = Column
*
*      | C0| C1| C2| C3| C4| C5| C6| C7| C8| C9| C10| C11| C12| C13| C14| C15
*    R0|0,0|0,1|0,2|0,3|0,4|0,5|0,6|0,7|0,8|0,9|0,10|0,11|0,12|0,13|0,14|0,15
*    R1|1,0|1,1|1,2|1,3|1,4|1,5|1,6|1,7|1,8|1,9|1,10|1,11|1,12|1,13|1,14|1,15
*
*                Example:
*                        To start the display on the 2nd row and 4th column,
*                        the following function call would be used.
*                        lcd_initialize (1,3);
*
* Return:      None
*
* Notes:       Additional explanation regarding the code
*
*      Note 1: When using an 8-bit interface, the function set command is sent
*              2 times to start communication with the LCD.
*
*              When sending this initial instruction, there must be a delay
*              between repeating the instruction. The minimum time for the
*              delay is defined in the datasheet. For purposes of this
*              program, a delay of 5 milliseconds was selected, which is
*              greater than the minimum amount of time required.
*
*              The reason a delay is programmed in instead of checking the
*              LCD busy flag is because according to the datasheet, the busy
*              flag will not be available initially.
*
*              Therefore, __delay_cycles() is used. Since the clock speed is
*              16 MHz, each clock cycle takes 0.0625 microseconds. To achieve
*              a delay of 5 milliseconds, 80000 cycles are needed.
*              See the description of __delay_cycles() in this file for detail
*              on how to do the calculations.
*
*              This instruction tells the LCD to use 8-bit mode, 2 lines
*              (0b00111000). Note when calling the function, the second
*              argument is a 0. This tells the compiler not to check the busy
*              flag.
*
*              When using the HD44780U chip, check its datasheet to determine
*              how many times this initial command must be executed and the
*              minimum time required for the delays.
    *********************************************************************************/
void lcd_initialize(unsigned int row, unsigned int column)
{
    /*Designate ports and pins used for I/O. */
    initialize_ports();

    /* Per datasheet, before initializing the LCD, it needs at least a
     * 40 millisecond delay. 50 milliseconds was chosen for this project. */
    __delay_cycles(800000);

    /* First time executing function set (see Note 1 in function description)*/
    send_instruction_to_lcd(0b00111000, 0);
    __delay_cycles(80000);                      /* delay 5 milliseconds */

    /* Second time executing function set (see Note 1 in function description)*/
    send_instruction_to_lcd(0b00111000, 0);
```

```
    __delay_cycles(80000);                          /* delay 5 milliseconds */

    send_instruction_to_lcd(0x0C, 1); /* Display on, cursor off, cursor blinking
                                         off */

    send_instruction_to_lcd(0x01, 1); /* Clear display */

    send_instruction_to_lcd(0x06, 1); /* Next character to the right */

    set_DDRAM_address(row, column); /* Set row/column where cursor starts */

}

/********************************************************************************
 * Name: initialize_ports();
 *
 * Description:This function initializes all the ports.
 *             - P1SEL and P2SEL designate Ports as I/O ports
 *             - P1DIR and P2DIR designate which pins are input or output pins.
 *               If a pin is set LOW (0), then it is an INPUT pin. If a pin is
 *               set HIGH (1), then it is a OUTPUT pin.
 *             - P1OUT and P2OUT set the pins HIGH or LOW.
 * Parameter:  None
 * Return:     None
 ********************************************************************************/
void initialize_ports()
{
    /* Port 1 set up */
    P1SEL = 0x00;  /* Set Port 1 to all I/O pins */
    P1DIR = 0x34;  /* 3 Bits are OUTPUT pins (P1.2, P1.4, and P1.5) */
    P1OUT = 0x00;  /* All pins are initially set LOW */

    /* Port 2 set up */
    P2SEL = 0x00;  /* Set Port 2 to all I/O pins */
    P2DIR = 0xFF;  /* Sets 8 Bits as OUTPUT (P2.0 - P2.7) */
    P2OUT = 0x00;  /* All pins are initially set LOW */
}

/********************************************************************************
 * Name: void check_busy_flag()
 *
 * Description: The ST7066U datasheet states that while the LCD is processing,
 *              an additional instruction cannot be accepted. The busy flag
 *              will remain HIGH (1) until the LCD is able to receive the
 *              next instruction. The busy flag is read through LCD pin #14
 *              (D7) when RS is LOW (0) and READ/WRITE is HIGH (1).
 *
 *              This function, checks the status of the busy flag and will
 *              return when the busy flag is LOW (0) indicating that the LCD
 *              is ready to receive a new instruction.
 *
 * Parameter:  None
 *
 * Return:     None
 *
 * Reference:  Newhaven NHD-0216K1Z datasheet (www.newhavendisplay.com)
 *
 * Notes:      Additional Explanation regarding code
 *
```

```
*      Note 1: The timing sequence in the datasheet was used as a guide for
*              checking the busy flag.
*
*              The steps are: (1) ENABLE_HIGH, (2) delay, (3) check busy flag
*              status, (4) delay, (5) ENABLE_LOW, (6) delay.
*
*              The ST7066U datasheet cautions:
*
*              Wait a minimum of 80 microseconds before checking the status of
*              the busy flag and do not keep ENABLE at HIGH continuously.
*              100 microseconds was selected for the delay.
*
*      Note 2: BIT7 is defined in MSP430.h as 0b10000000. Bit 7 of P2IN holds
*              the status of the busy flag (1 for busy, and 0 for not busy).
*              If the busy flag is 1, then when P2IN is AND'ed with BIT7,
*              the variable busy_flag will not be 0 and the loop will continue.
*              If the busy flag is 0, then when P2IN is AND'ed with BIT7,
*              the variable busy_flag will be assigned the value of 0.
*              This will cause the loop to exit.
*   ***************************************************************************/
void check_busy_flag()
{

    COMMAND_MODE;          /* Set pin P1.4 to 0 --> this is the RS pin */
    READ_MODE;             /* Set pin P1.2 to 1 --> this is the R/W pin */

    P2DIR &= ~(BIT7);      /* Makes P2.7 an input pin which will be used to read
                              the busy flag status of the LCD. */

    P2REN |= BIT7;         /* Enables pull-up/pull-down resistors on P2.7 */

    P2OUT |= BIT7;         /* Selects P2.7 to engage pull-up resistor
                              (1 = pull-up and 0 = pull-down ) */

    int busy_flag;         /* Variable to hold the status of the LCD busy flag */

    do                     /* Loop checks the busy flag status */
    {

        /* See Note 1 in the function description */

        ENABLE_PIN_HIGH;            /* ENABLE is set HIGH */

        __delay_cycles(1600);       /* Delay 100 microseconds */

        busy_flag = (P2IN & BIT7); /* See Note 2 in the function description */

        ENABLE_PIN_LOW;       /* ENABLE is set LOW.*/

        __delay_cycles(1600); /* Delay 100 microseconds */

    } while (busy_flag); /* Exit loop when busy_flag = 0 */

    P2DIR |= BIT7;         /* Makes P2.7 a 1; which makes P2.7 an output pin */
    P2REN &= ~(BIT7);      /* Disengage the pull-up/pull-down resistors on P2.7 */
    P2DIR = 0xFF;          /* Make P2.0 - P2.7 output pins */
}


/*****************************************************************************
```

```
* Name: void set_DDRAM_address(unsigned int row, unsigned int column)
*
* Description: Purpose is to set the cursor of the LCD to the desired row and
*              column that the cursor will begin writing.
*
*                 - The DDRAM (Display Data RAM) has a base address number of
*                   0b10000000 (decimal 128). The remaining bits of the address
*                   are based on the columns (0-15).
*
*                 - This means that to find the DDRAM address to start in the
*                   first row (row 0), add 128 to the column number.
*
*                 - To find the DDRAM address for the second row, consider the
*                   following:
*
*                   The DDRAM has an extended capability to display up to 40
*                   characters per row. The 16x2 LCD does not have this
*                   extended capability -- it can only display 16 characters per
*                   row.
*
*                   For the 16x2 LCD Display, the Driver Chip starts the second
*                   row address at the base address (128 + 64 = 192).
*
*                   This means that to find the DDRAM address to start in the
*                   second row (row 1), add 192 to the column number.
*
*                   For more detail regarding how the DDRAM works, please see the
*                   referenced article below and the datasheet.
*
*
* Parameters: Two unsigned integers. The first represents the row that the
*             display will start at, and the second unsigned integer represents
*             the column. The LCD has 2 rows and 16 columns. The rows are
*             numbered 0 and 1 and the columns are numbered 0 - 15.
*
*             Please see the table in the description of the function
*             lcd_initialization() above which shows the row/column
*             numbering.
*
* Return:     None
*
* References: "Controlling a 16x2 LCD Display with the MSP430G2553
*             Microcontroller Using Assembly Code" in Appendix.
*
*             Newhaven NHD-0216K1Z datasheet (www.newhavendisplay.com)
******************************************************************************/

void set_DDRAM_address(unsigned int row, unsigned int column)
{
    unsigned int ddram_addr;                      /*Variable to hold the address */

    if(row == 0)                                  /* If address is in first row */
    {
        ddram_addr = 128 + column;                /* Add 128 and column number */
        send_instruction_to_lcd(ddram_addr, 1);   /* Check busy flag, then send
                                                     the command to the LCD */
    }

    if(row == 1)                                  /* If address is in second row */
```

```
    {
        ddram_addr = 192 + column;              /* Add 192 and column number */
        send_instruction_to_lcd(ddram_addr, 1); /* Check busy flag, then send
                                                    the command to the LCD */
    }
}


/*******************************************************************************
 * Name: void send_instruction_to_lcd(unsigned int instruction, unsigned int check_busy)
 *
 * Description: Checks the busy flag, if indicated, then sends an instruction
 *              to the LCD. These commands are used to initialize the LCD and
 *              configure settings.
 *
 * Parameters:  Takes two unsigned integers. The first integer represents the
 *              instruction. The second integer represents if the busy flag for
 *              the LCD should be checked first.
 *
 *              Use a 1 for the second parameter to indicate the busy flag
 *              should be checked prior to sending the instruction.
 *
 *              Use a 0 for the second parameter to indicate the busy flag
 *              should not be checked prior to sending the instruction.
 *
 *              This second parameter accommodates the situation when the LCD
 *              is first started up. According to the datasheet, when the
 *              initial commands are sent to the LCD, the busy flag is not
 *              available yet.
 *
 * Return:      None
 *
 * Reference:   Newhaven NHD-0216K1Z datasheet (www.newhavendisplay.com)
 *
 *              "Controlling a 16x2 LCD Display with the MSP430G2553
 *              Microcontroller Using Assembly Code" in Appendix
 *
 * Notes:       Additional explanation regarding the code
 *
 *      Note 1: The ST7066U datasheet specifies the minimum delay required after
 *              ENABLE is brought both HIGH and LOW. For purposes of this code
 *              and consistency, a delay of 100 microseconds was selected.
 *******************************************************************************/

void send_instruction_to_lcd (unsigned int instruction, unsigned
int check_busy) {
    if(check_busy)
    {
        check_busy_flag(); // Check if LCD is ready to accept a new instruction
    }

    WRITE_MODE;       /* Set to WRITE_MODE. Pin P1.2 is LOW (0) */
    COMMAND_MODE;     /* Set to COMMAND MODE. Pin P1.4 is LOW (0) */
    ENABLE_PIN_HIGH;  /* Bring Pin P1.5 HIGH (1) */

    /* See Note 1 in Function Description */

    __delay_cycles (1600); /* Delay 100 microseconds */

    P2OUT = instruction;  /* Load instruction into P2OUT which sends it to
```

```
                            the LCD */

    __delay_cycles (1600); /* Delay to allow LCD to process instruction */

    ENABLE_PIN_LOW;          /* Bring Pin P1.5 LOW (0) */
    __delay_cycles (1600);   /* Delay 100 microseconds */
}

/**************************************************************************
 * Name: void write_ASCII_to_lcd(char data)
 *
 * Description: Function sends data to the LCD to be displayed on the screen
 *
 * Parameter:   Takes a char which represents an ASCII character/value
 *
 * Return:      None
 **************************************************************************/

void write_ASCII_to_lcd(char ASCII)
{
    check_busy_flag();                 /* Check if LCD is ready to accept data */

    WRITE_MODE;                        /* To write to LCD, R/W is put to LOW (0) */

    DATA_MODE;                         /* Set RS to HIGH(1) for data */

    ENABLE_PIN_HIGH;                   /* Set ENABLE to HIGH (1) */
    __delay_cycles(1600);              /* Delay for 100 microseconds */

    P2OUT = ASCII;                     /* Load data into P2OUT -> send to LCD. */
    __delay_cycles(1600);              /* Delay for 100 microseconds */

    ENABLE_PIN_LOW;                    /* Put ENABLE to LOW (0) */
    __delay_cycles(1600);              /* Delay for 100 microseconds */
}

/**************************************************************************
 * Name: void display_info(char* information)
 *
 * Description: Function iterates through the characters of a C-string (an array
 *              of chars) until it reaches the NULL terminator.
 *
 *              - During the iteration, it sends the current ASCII character to
 *                be written to the LCD.
 *              - Then it moves to the next character by incrementing to the
 *                next index until it reaches the NULL terminator.
 *
 * Parameter:   Takes a pointer to a C-String
 *
 * Return:      None
 **************************************************************************/
void display_info(char* information)
{
    unsigned int i = 0;

    /* Continue loop until reaching NULL terminator */
    while (information[i] != NULL)
    {
        /* Send current character being pointed to to be printed to LCD */
```

```
        write_ASCII_to_lcd(information[i]);

        /* Increment to next character to be read */
        i++;
    }
}


/*********************************************************************************
 * Name: ADC_initialize()
 *
 * Description: This function initializes the peripheral module,
 *              Analog-to-Digital Converter (ADC).
 *
 *              - This program uses ADC Channel 7 (P1.7)
 *
 * Parameter:  None
 *
 * Return:     None
 *
 * Reference: Texas Instruments SLAU144K Users Guide (www.ti.com)
 *
 * Notes:     Additional explanation regarding the code
 *
 *     Note 1: msp430.h defines BIT7 as 0x0080 or 0b10000000.
 *             When the value of BIT7 is assigned to ADC10AE0, BIT7 corresponds
 *             to the Analog Input A7 which is pin P1.7. This enables Channel 7 of
 *             the ADC.
 *********************************************************************************/
void ADC_initialize()
{
    ADC10CTL0 = ADC10CTL0 | ADC10SHT_2 | ADC10ON; /* Setting up ADC */
    ADC10CTL1 = ADC10CTL1 | INCH_7;                /* Setting up ADC on Channel 7 */

    ADC10AE0 = BIT7; /* See Note 1 in the Function description */
}


/*********************************************************************************
 * Name: const char* integer_to_ASCII(char ascii[], unsigned int value, unsigned int length)
 *
 * Description: Purpose is to convert base 10 integers (values) to ASCII
 *              characters that are added to a C-String (array of chars that are
 *              NULL terminated). This C-String will be returned and sent to be
 *              displayed on the LCD screen.
 *
 *              Description of Logic Used:
 *
 *              - First find the number of digits in the value.
 *
 *              - Once the number of digits in the value is determined, the
 *                number of indices in the array needed to display the ASCII
 *                characters will be known. Any extra indices before the NULL
 *                terminator (last index of array), will be replaced with
 *                "blank spaces" so that if the previous value had more digits
 *                than the current value, the extra digits on the LCD display
 *                will be overwritten with blank spaces and no longer displayed.
 *
 *                Note: replacing extra indices with a blank avoids having to
 *                clear the LCD screen each time a value changes and minimizes
 *                flicker.
```

```
*
*        - A loop is used to iterate through the indices of the
*          array. A variable called index is initialized to 0 and
*          incremented with each iteration of the loop. It is used to
*          track the current index of the array. This loop will
*          continue as long as the index is less than the (length of the
*          array - 1). The last index of the array is reserved for the
*          Null terminator.
*
*          Within the loop, conditional statements are used to check
*          if the index is less than the number of digits in the
*          value.
*
*            - If the index is less than the number of digits, take the
*              following steps:
*
*                - Find the modulo (remainder) and quotient of the value/10.
*                - Update the value to the quotient.
*
*                  This repeated integer division and finding the modulo,
*                  parses out each digit starting at the right-most digit of
*                  the integer and going toward the left-most digit.
*                  This right to left parsing of the digits of the integer
*                  occurs for the following reasons:
*
*                  When an integer is repetitively divided by 10, until the
*                  quotient is 0, the first division's modulo gives the
*                  number of 10^0 power. 10^0 power is the right-most digit
*                  of a base 10 integer. Each subsequent division yields
*                  progressively higher orders of powers of 10 (10^1, 10^2...)
*                  depending on the number of digits. Each of these powers of
*                  10, are placed to the left of the previous digit.
*
*                  Example: using the number 84:
*                  The first division by 10 gives a quotient of 8 and a modulo
*                  of 4. The second division by 10 gives a quotient of 0 and a
*                  modulo of 8. No further division is necessary because the
*                  second division yielded a quotient of 0.
*                  This means that 84 = (8 * 10^1) + (4 * 10^ 0).
*
*                - To complete the conversion to an ASCII value that represents
*                  numerals, add 48 to each digit. This is because the ASCII
*                  decimal values for numerals begin at 48.
*
*                - Place the calculated ASCII value in the array.
*
*                  In order to place the ASCII values from right to left, use
*                  a second variable that represents the index of the array
*                  where the current digit (ASCII value) will be placed. In
*                  this program, this variable is called "current_digit".
*                  Variable "current_digit" is initialized to 1 and will be
*                  incremented with each iteration.
*
*                  Subtract current_digit from the number of digits to
*                  get the index of the array where the calculated ASCII
*                  value will be placed. This results in a right to left
*                  placement of the ASCII characters in the array instead of
*                  the usual order of left to right placement.
*
```

```
*                    - If the starting index is not less than the number
*                      of digits but is less than the (length of the array - 1),
*                      take the following steps:
*
*                        - Assign the ASCII value 32 at the current index of the
*                          array. The ASCII value 32 is a blank space.
*
*                    - The last index of the array already has a NULL terminator in
*                      it because the C-String was cleared at the beginning of the
*                      function.
*
* Parameters: Takes an array of chars that is used to create the C-String, an
*             unsigned integer value, and the length of the C-String to be
*             created. This length is defined as a constant (SIZE) in main.c
*
* Return:     A pointer to an array of chars.
*
* Reference: B. Kernighan; D. Ritchie, The C Programming Language, 2nd ed.,
*            Prentice-Hall Software Series, 1988, pgs. 60 - 64.
********************************************************************************/
const char* integer_to_ASCII(char ascii[], unsigned int value,
unsigned int length) {
    /* Clear the C-String passed in of previous values by setting each index
     * to NULL. The length of the C-String is defined in main.c as SIZE */
    clearString(ascii, length);

    /* Declare a variable to hold the number of digits in the value */
    unsigned int number_digits = 0;

    /* Get the number of digits in the value -> gives number of indices
     * required in the C-String to represent the value */
    number_digits = number_of_digits(value);

    /* Declare variables used to hold results of the modulo and current_digit
     * described in the function description.*/
    unsigned int modulo, current_digit = 1;
    unsigned int index = 0;   /* initialized to first index of array */

    /* Use a loop to iterate through the array*/
    while(index < length - 1)
    {
        if(index < number_digits)
        {
            modulo = get_modulo(value, 10); /* Get remainder of value/10 */
            value = division(value, 10);     /* Get quotient of value/10 */

            /* Add calculated ASCII value to C-String from R -> L */
            ascii[number_digits - current_digit] = modulo + 48;
        }

        else
        {
            /* If the index is >= number_digits, place blank space at the
               current index */
            ascii[index] = 32;
        }

        index++;                /* Increment the index */
        current_digit++;        /* Increment current_digit */
```

```
    }

    return ascii;
}

/********************************************************************************
 * Name: const char * clearString(char string[], unsigned int length)
 *
 * Description: This function iterates through a C-String and replaces all the
 *              index values with '\0' or NULL. It "clears" the C-String
 *
 * Parameter:   An array of chars and an integer that defines the length.
 *
 * Return:      A pointer to a C-String
 ********************************************************************************/
const char * clearString(char string [], unsigned int length)
{
    unsigned int i;             /* Declare i to use for the index */
    for(i = length; i > 0; i--)  /* Iterate through the array of chars */
    {
        string[i - 1] = NULL;  /* Set the value at each index to NULL or '\0' */
    }
    return string;

}


/********************************************************************************
 * Name: unsigned int number_of_digits(unsigned int value)
 *
 * Description:Purpose is to find the number of digits in a base 10 unsigned
 *              integer.
 *
 *              The digits in a base 10 integer represent powers of 10.
 *              There are different approaches that can be used to find the
 *              number of digits in the value.
 *
 *              This function uses integer division by 10 and repeatedly divides
 *              an unsigned base 10 integer by 10. With each division it counts
 *              the number of times division occurs and stops when the quotient
 *              is no longer >= 1. The number of times division occurs represents
 *              the largest power of ten -1 in the value. Each one of those powers
 *              of 10 is represented by a single digit.
 *
 *              If a number can only be divided once and the count representing
 *              the number of digits is 1, then the largest power of 10 in the
 *              number is 10^0.
 *              If the number can be divided twice, then the number of digits is
 *              2 and the largest power of 10 in number is 10^1.
 *              If the number can be divided three times, then the number of
 *              digits is 3 and the largest power of 10 in the number is 10^2 ...
 *              etc.
 *
 *              Example: 801
 *
 *              Using integer division which rounds the quotient down to the
 *              closest integer, the number 801 will be divided 3 times by 10
 *              before the quotient is no longer >= 1. The number of divisions
 *              is 3 and the largest power of 10 in 801 is 10^2.
 *
```

```
*               Note: the exception used in this function is when the
*               integer value is 0. For this case, a 1 is returned.
*
* Parameter:  Takes an unsigned integer which represents the input value
*
* Return:     An integer that represents the number of digits in the value.
*
* For further understanding, read about positional notation and place value.
******************************************************************************/
unsigned int number_of_digits(unsigned int value)
    {
    /* Initialize the number of digits to 0 */
    unsigned int number_digits = 0;

    /* Check if the input value is 0 */
    if(value == 0)
    {
        number_digits = 1;
    }

    /* Input value != 0  -> begin integer division */
    else
    {
        /* Continue integer division by 10 until value is not >= 1
         * Increment the number of digits with each iteration of loop. */
        while(value >= 1)
        {
            value = division(value, 10); /* Call division function to divide
                                            by 10 */
            number_digits++;
        }
    }

    /* Return the number of digits counted. */
    return number_digits;
}


/******************************************************************************
 * Name: unsigned int division(unsigned int dividend, unsigned int divisor)
 *
 * Description:Purpose is to perform unsigned integer division using
 *             subtraction.
 *
 *             STEPS:
 *             - Check if the dividend is less than the divisor. If it is,
 *               then return 0.
 *             - Check if the divisor is 0. If it is, display error message.
 *               Then return 65535 to signify an error because division by
 *               0 is undefined. The value of 65535 was chosen as an error
 *               message because it is the largest unsigned integer that
 *               can be held in a 16-bit register.
 *             - Loop:
 *               As long as the dividend is >= to the divisor, subtract
 *               the divisor from the dividend and increment the quotient
 *             - Return the quotient.
 *
 * Parameters: Two unsigned integers: the dividend and divisor.
 *
 * Return:     An integer that represents the quotient unless division by 0
```

```
*              is attempted. This will display error message and return 65535.
*
* NOTE:        Bit-wise operations can also be used to do division without
*              using the division operator. An Internet search will reveal
*              options to use. If speed is critical, you may wish to time
*              both options and determine which is faster.
*
* Reference: T. Floyd, Digital Fundamentals, 9th ed., Prentice-Hall, 2006, pgs. 73-74.
***********************************************************************/
unsigned int division(unsigned int dividend, unsigned int divisor)
{
    unsigned int quo = 0;              /* Variable holds quotient */
    if (dividend < divisor)            /* Check if divisor is < dividend */
    {
        return 0;
    }

    if (divisor == 0)                  /* Check if divisor is 0. */
    {
    send_instruction_to_lcd(0x01, 1); /* Clear display */
    display_info("Can't div by 0");    /* Show error message */
    set_DDRAM_address(1,0);
    display_info("Next value bad");
    __delay_cycles(80000000);              /* Pause 5 seconds */
    send_instruction_to_lcd(0x01,1);       /* Clear display */
    display_info(" ADC OUTPUT");           /* Restore original info */
        return 65535;                  /* Return "error" value if divisor = 0 */
    }

    while (dividend >= divisor)        /* Loop until dividend < divisor */
    {
        dividend = dividend - divisor; /* Subtract divisor from dividend */
        quo++;                         /* Increment the quotient */
    }
    return quo;

}


/********************************************************************************
* Name: unsigned int get_modulo(unsigned int dividend, unsigned int divisor)

* Description: Purpose is to find the modulo (remainder after division)
*              without using the modulus operator.
*
*              - To check the correctness of integer division, the equation
*                below is used.
*
*                      (quotient x divisor) + modulo = dividend.
*
*              Solving the equation for modulo:
*
*                      modulo = dividend - (quotient x divisor).
*
*              Use this formula to find the modulo without using the
*              modulus operator.
*
* Parameters: Takes two unsigned integers. The first is the dividend and
*             the second is the divisor.
*
```

```
* Return:     An integer that represents the modulo.
*
* NOTE:       This function will not work correctly if the divisor is 0.
*             Division by 0 is not defined. If this case occurs, the function
*             returns the value of the dividend.
*****************************************************************************/
unsigned int get_modulo(unsigned int dividend, unsigned int
divisor)
{

    unsigned int quo = division(dividend, divisor); /* Get the quotient */
    unsigned int quo_x_divisor = multiply(quo, divisor);  /* Get quotient x divisor */

    /* Calculate the modulo using the formula:
     * modulo = dividend - (quo x divisor); then, return this value. */
    return (dividend - quo_x_divisor);
}


/*****************************************************************************
* Name: unsigned int multiply(unsigned int multiplicand, unsigned int multiplier)
*
* Description: Purpose is to find the product of two unsigned integers without
*             using the multiplication operator.
*
*             The logic uses the concept that multiplication adds the
*             value of the multiplicand to itself N times (where N is the
*             multiplier). This cumulative sum produces the product.
*
*             Note: bitwise operators can also be used to find the product
*             without using the multiplication operator. An Internet search
*             will reveal options to use. If speed is critical, you may wish
*             to time both options and determine which is faster.
*
* Parameters: Takes two unsigned integers. The first is the multiplicand and
*             the second is the multiplier
*
* Return:     An unsigned integer that represents the product
*
* Reference:  T. Floyd, Digital Fundamentals, 9th ed., Prentice-Hall, 2006, pgs. 70-72.
*****************************************************************************/
unsigned int multiply(unsigned int multiplicand, unsigned int multiplier)
{
    unsigned int product = 0; /* Declare/initialize variable to store product*/

    int i;                    /* Declare "i" to use to iterate thru the loop*/

    for (i = multiplier; i > 0; i--)
    {
        product = product + multiplicand; /* add multiplicand to itself */
    }

    return product;
}


/*****************************************************************************/
```

Appendix G
How to Add Files into Code Composer

The example code EXAMPLE_LCD_ADC has 3 files[3]:

- main.c: contains main
- msp430_lcd_adc.h: contains MACRO definitions and function declarations
- msp430_lcd_adc.c: contains an explanation of how to use __delay_cycles() and function definitions

1. Open Code Composer and follow the instructions on how to create a new project as described in Lab 2. Give a name to the project.
2. The project will open with an empty default main.c file.
3. Open the Project Explorer Window by going to View→Project Explorer (see Fig. G.1).
4. A new Window will open (see Fig. G.2). The new empty project just created is listed: Example_LCD_ADC [Active–Debug]. Click on the " > " to the left of the project name to expand the contents of the project.
5. The expanded contents will be similar folders/files in Fig. G.3.
6. Two files, msp430_lcd_adc.c and msp430_lcd_adc.h need to be added. For simplicity, the files will all be placed in the Example_LCD_ADC project directory.
7. Right click on the name of the project, Example_LCD_ADC (the project may have been named differently). Menus appear as in Fig. G.4, select "New" and click on "File."
8. Selecting "File" opens a new window (see Fig. G.5) where the two new files will be added to the directory. The parent folder is the name of the project (Example_LCD_ADC). For the Files names, use the exact name of the provided code. Add the C code file (msp430_lcd_adc.c)

---

[3] This Appendix was written by D. Kretzschmar.

© The Editor(s) (if applicable) and The Author(s), under exclusive license to Springer Nature Switzerland AG 2023
J. Kretzschmar et al., *MSP430 Microcontroller Lab Manual*, Synthesis Lectures on Digital Circuits & Systems, https://doi.org/10.1007/978-3-031-26643-0

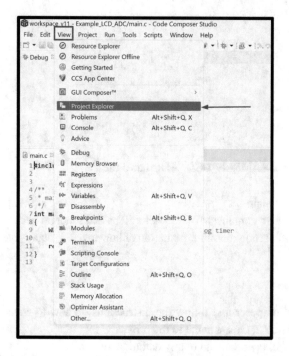

Fig. G.1 New project

9. Repeat this procedure to add the second file (msp430_lcd_adc.h). Once you have added the two new files, the Project Explorer Window should look similar to the image in Fig. G.6.

10. For each of the three files, copy/paste the associated code into it.
    Note: When copying code from a pdf it is easy to miss a closing brace or semi–colon from the code. It is also easy to accidently include page titles, or other unintended material. These copy errors can be the reason the code may not build or run.

11. Save the code, build, and run the code as described in Lab 2.

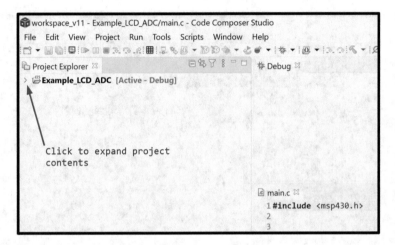

Fig. G.2 New empty project

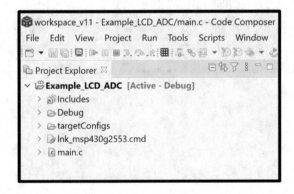

Fig. G.3 Expanded contents

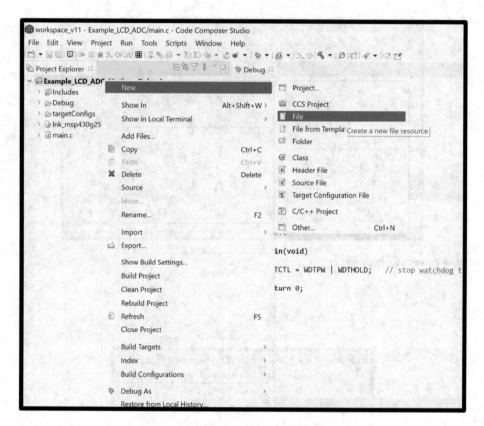

Fig. G.4 Adding files

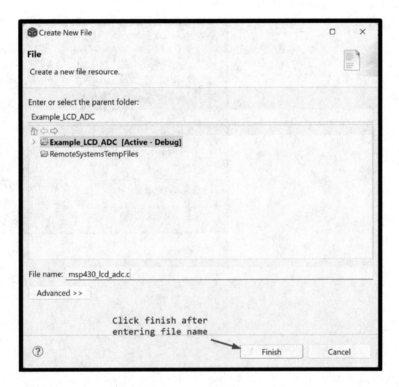

Fig. G.5 Added files

Fig. G.6 Adding files

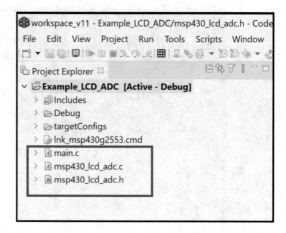

Appendix H
Tidbits–Useful Miscellaneous Information

H.1 Introduction

In the course of preparing these exercises for the labs several bits and pieces of information were "discovered." Some of these glitches took several hours to discover and figure out. This is a compilation of tidbits of information that may save from spending several hours trying to figure out why something isn't working out as planned. This miscellaneous information is presented in no particular order.

1. In assembly code, the statement:

```
mov.b   #BIT0|BIT6, &P1OUT
```

tells the microcontroller to make pins P1.0 and P1.6 on Port 1 HIGH (3.3 V on the pins). Alternative ways to write this command are:

```
mov.b      #0x41, &P1OUT          (Hex notation)
mov.b      #01000001b, &P1OUT     (Binary notation)
mov.b      #65, &P1OUT            (Decimal notation)
```

Another example:

```
mov.b      #BIT1|BIT5, &P1OUT

mov.b      #0x22, &P1OUT          (Hex notation)
mov.b      #00100010b, &P1OUT     (Binary notation)
mov.b      #34, &P1OUT            (Decimal notation)
```

© The Editor(s) (if applicable) and The Author(s), under exclusive license to Springer Nature Switzerland AG 2023
J. Kretzschmar et al., *MSP430 Microcontroller Lab Manual*, Synthesis Lectures on Digital Circuits & Systems, https://doi.org/10.1007/978-3-031-26643-0

Fig. H.1 Label alignment

```
18 ;--------------------------------------------------------
19 RESET        mov.w   #__STACK_END,SP          ; Initialize
20 StopWDT      mov.w   #WDTPW|WDTHOLD,&WDTCTL   ; Stop watchd
21
22
23 ;--------------------------------------------------------
24 ; Main loop here
25 ;--------------------------------------------------------
26              mov.b   #BIT0, &P1DIR
27 OSCILLATE:
28              bis.b   #BIT0, &P1OUT
29              nop
30              nop
31              nop
32              bic.b   #BIT0, &P1OUT
33              nop
34              nop
35              nop
36              jmp     OSCILLATE
37
38
39 ;--------------------------------------------------------
40 ; Stack Pointer definition
41 ;--------------------------------------------------------
42              .global __STACK_END
43              .sect   .stack
44
```

2. The label "OSCILLATE" on line 27 in Fig. H.1 needs to be aligned on the far left (no indentation), or the program will not compile and produce an error message. All labels need to follow this format.

3. When first setting up the Code Composer IDE there is a selection of where code projects will be saved. The default is on the hard drive of the computer in a folder named "work-place_v11" (version 11 is the latest available). Within this folder program folders/files can be located and the program itself can be found in the file that says "main." The "main" file will have "C File" or "ASM File" noted at the end of the line. Earlier versions such as "workplace_v9," or "workplace_v10" are difficult to directly load into the latest version of Code Composer and require a copy and paste of the program approach. If using a text editor such as Notepad++ saving programs with the appropriate extension (i.e., ".c" or ".asm" may preclude any issues when transferring into Code Composer.

4. How to set up the VLOCLK (very low frequency oscillator). In the MSP430 Users Guide (Texas Instruments Document SLAU144K) p. 279, Sect. 5.2.2 "VLOCLK source is selected by setting $LFXT1Sx = 10$ when $XTS = 0$." LFXT1Sx is in (Register BCSCTL3 Bits 5–4) and XTS is in (Register BCSCTL1 Bit 6). With those two parameters set, the VLOCLK can be selected in (Register BCSCTL2 Bits 7–6 = "11").

5. How to use pins P2.6 and P2.7 as I/O pins. In the MSP430 Users Guide (Texas Instruments Document SLAU144K) p. 349, Sect. 8.3.7 register &P2SEL is set to #11000000b (#C0h in hexadecimal notation) when power is applied to the microcontroller. For P2.6 and P2.7 to be used as I/O pins then register &P2SEL needs to be changed to #00000000b (#00h hexadecimal notation).

6. The assembly instruction "dadd" converts a binary number to a BCD number and is only available when programming in assembly (not available when programming in C.

7. Be careful, sometimes you will get so used to writing code with statements such as:

```
mov     #45, R8
```

that when you want to move a register (such as the output from the ADC or TIMER) you may accidently make the mistake of writing the code using the pound sign instead of the ampersand symbol:

```
mov     #TA0CCR0, R8
```

It will compile, however, the program will not work. The statement must be written as:

```
mov     &TA0CCR0, R8
```

8. Be careful when working with the 16–bit registers of the MSP430. A statement such as "mov.b" is not the same as "mov.w". "mov.w" indicates you are working with all 16 Bits of the register. Using "mov.b" means that only the lower 8 Bits of a register are recognized (such as an address location). The program will compile; however, the result may not be what is expected. For example:

```
mov.b   #8793, R12
```

Register 12 will only have 89 (decimal) or 01011001 (binary). Only the lower 8 bits of 8793 (decimal) or 0010 0010 0101 1001 (binary) will be loaded into R12.

Appendix I
Operators in C

There are many operators provided in the C language. An abbreviated list of common operators are provided in Figs. I.1 and I.2. The operators have been grouped by general category. The symbol, precedence, and brief description of each operator are provided. The precedence column indicates the priority of the operator in a program statement containing multiple operators. Only the fundamental operators are provided.[4]

I.1   General Operations

Within the general operations category are brackets, parenthesis, and the assignment operator. We have seen in an earlier example how bracket pairs are used to indicate the beginning and end of the main program or a function. They are also used to group statements in programming constructs and decision processing constructs. This is discussed in the next several sections.

The parenthesis is used to boost the priority of an operator. For example, in the mathematical expression $7 \times 3 + 10$, the multiplication operation is performed before the addition since it has a higher precedence. Parenthesis may be used to boost the precedence of the addition operation. If we contain the addition operation within parenthesis, $7 \times (3 + 10)$, the addition will be performed before the multiplication operation and yield a different result from the earlier expression.

The assignment operator ($=$) is used to assign the argument(s) on the right–hand side of an equation to the left–hand side variable. It is important to insure the left and the right–hand side of the equation have the same type of arguments. If not, unpredictable results may occur.

---

[4] This section was adapted with permission from Arduino II: Systems, S.F. Barrett, Morgan & Claypool Publishers, 2020.

© The Editor(s) (if applicable) and The Author(s), under exclusive license to Springer Nature Switzerland AG 2023

J. Kretzschmar et al., *MSP430 Microcontroller Lab Manual*, Synthesis Lectures on Digital Circuits & Systems, https://doi.org/10.1007/978-3-031-26643-0

General		
Symbol	Precedence	Description
{ }	1	Brackets, used to group program statements
( )	1	Parenthesis, used to establish precedence
=	12	Assignment

Arithmetic Operations		
Symbol	Precedence	Description
*	3	Multiplication
/	3	Division
+	4	Addition
-	4	Subtraction

Logical Operations		
Symbol	Precedence	Description
<	6	Less than
<=	6	Less than or equal to
>	6	Greater
>=	6	Greater than or equal to
==	7	Equal to
!=	7	Not equal to
&&	9	Logical AND
‖	10	Logical OR

Fig. I.1 C operators. (Adapted from [Barrett and Pack])

I.2   Arithmetic Operations

The arithmetic operations provide for basic math operations using the various variables described in the previous section. As described in the previous section, the assignment operator ($=$) is used to assign the argument(s) on the right–hand side of an equation to the left–hand side variable.

Example: In this example, a function returns the sum of two unsigned int variables passed to the function.

Bit Manipulation Operations		
Symbol	Precedence	Description
<<	5	Shift left
>>	5	Shift right
&	8	Bitwise AND
^	8	Bitwise exclusive OR
|	8	Bitwise OR

Unary Operations		
Symbol	Precedence	Description
!	2	Unary negative
~	2	One's complement (bit-by-bit inversion)
++	2	Increment
--	2	Decrement
type(argument)	2	Casting operator (data type conversion)

Fig. I.2 C operators (continued). (Adapted from [Barrett and Pack])

```
unsigned int  sum_two(unsigned int variable1, unsigned int variable2)
{
unsigned int  sum;

sum = variable1 + variable2;

return sum;
}
```

I.3   Logical Operations

The logical operators provide Boolean logic operations. They can be viewed as comparison operators. One argument is compared against another using the logical operator provided. The result is returned as a logic value of one (1, true, high) or zero (0, false, low). The logical operators are used extensively in program constructs and decision processing operations to be discussed in the next several sections.

I.4   Bit Manipulation Operations

There are two general types of operations in the bit manipulation category: shifting operations and bitwise operations. Let's examine several examples:

Example: Given the following code segment, what will the value of variable2 be after execution?

```
unsigned char    variable1 = 0x73;
unsigned char    variable2;

variable2 = variable1 << 2;
```

Answer: Variable "variable1" is declared as an eight–bit unsigned char and assigned the hexadecimal value of $(73)_{16}$. In binary notation this is expressed as $(0111\_0011)_2$. The $<<$ 2 operator provides a left shift of the argument by two places. After two left shifts of $(73)_{16}$, the result is $(cc)_{16}$ and will be assigned to the variable "variable2." Note that the left and right shift operation is equivalent to multiplying and dividing the variable by a power of two respectively.

The bitwise operators perform the desired operation on a bit–by–bit basis. That is, the least significant bit of the first argument is bit–wise operated with the least significant bit of the second argument and so on.

Example: Given the following code segment, what will the value of variable3 be after execution?

```
unsigned char    variable1 = 0x73;
unsigned char    variable2 = 0xfa;
unsigned char    variable3;

variable3 = variable1 & variable2;
```

Answer: Variable "variable1" is declared as an eight–bit unsigned char and assigned the hexadecimal value of $(73)_{16}$. In binary notation, this is expressed as $(0111\_0011)_2$. Variable "variable2" is declared as an eight–bit unsigned char and assigned the hexadecimal value of $(fa)_{16}$. In binary notation, this is expressed as $(1111\_1010)_2$. The bitwise AND operator is specified. After execution variable "variable3," declared as an eight–bit unsigned char, contains the hexadecimal value of $(72)_{16}$.

I.5   Unary Operations

The unary operators, as their name implies, require only a single argument. For example, in the following code segment, the value of the variable "i" is incremented. This is a shorthand method of executing the operation "$i = i + 1$;"

```
unsigned int    i;

i++;
```

Example: It is common in embedded system design projects to have every pin on a micro-controller employed. Furthermore, it is not uncommon to have multiple inputs and outputs assigned to the same port but on different port input/output pins. Some compilers support specific pin reference. Another technique that is not compiler specific is **bit twiddling**. Figure I.3 provides bit twiddling examples on how individual bits may be manipulated without affecting other bits using bitwise and unary operators. The information provided here was extracted from the ImageCraft JumpStart C for AVR compiler documentation [www. ImageCraft.com].

Syntax	Description	Example	
a | b	bitwise or	PORTB |= 0x80;	// turn on bit 7 (msb)
a & b	bitwise and	if ((PINB & 0x81) == 0)	// check bit 7 and bit 0
a ^ b	bitwise exclusive or	PORTB ^= 0x80;	// toggle/flip bit 7
~a	bitwise complement	PORTB &= ~0x80;	// turn off bit 7

Fig. I.3 Bit twiddling [ImageCraft]

Appendix J
UML Activity Diagram

The UML Activity Diagram is simply a UML compliant flow chart. UML is a standardized method of documenting systems. The activity diagram is one of the many tools available from UML to document system design and operation. The basic symbols used in a UML activity diagram for a microcontroller based system are provided in Fig. J.1 [Fowler].

To develop the UML activity diagram for a system, we can use a top–down, bottom–up, or a hybrid approach. In the top–down approach, we begin by modeling the overall flow of the algorithm from a high level.

You should work out all algorithm details at the UML activity diagram level prior to coding any software. If you can not explain system operation at this higher level, first, you have no business being down in the detail of developing the code. Therefore, the UML activity diagram should be of sufficient detail so you can code the algorithm directly from it.

Fig. J.1 UML activity diagram symbols. Adapted from [Barrett and Pack]

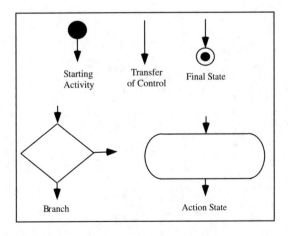

J. Kretzschmar et al., *MSP430 Microcontroller Lab Manual*, Synthesis Lectures on Digital Circuits & Systems, https://doi.org/10.1007/978-3-031-26643-0

J.1   References

- M. Fowler with K. Scott "UML Distilled – A Brief Guide to the Standard Object Modeling Language," 2nd edition, Boston:Addison–Wesley, 2000.

Printed in the United States
by Baker & Taylor Publisher Services